Interoperability in IoT for Smart Systems

Intelligent Systems
Series Editor: Prasant Kumar Pattnaik

Interoperability in IoT for Smart Systems
Monideepa Roy, Pushpendu Kar and Sujoy Datta

For more information about this series, please visit: https://www.routledge.com/Intelligent-Systems/book-series/IS

Interoperability in IoT for Smart Systems

Edited by

Monideepa Roy, Pushpendu Kar and Sujoy Datta

CRC Press is an imprint of the
Taylor & Francis Group, an **informa** business

First edition published 2021
by CRC Press
6000 Broken Sound Parkway NW, Suite 300, Boca Raton, FL 33487-2742
and by CRC Press
2 Park Square, Milton Park, Abingdon, Oxon OX14 4RN

ISBN: 978-0-367-51986-5 (hbk)

ISBN: 978-1-003-05597-6 (ebk)

Typeset in Times LT Std

by KnowledgeWorks Global Ltd.

Contents

Foreword vii
Preface ix
About the Authors xiii

Chapter 1 Internet of Things: Challenges and Its Applications 1
Zeenat Rehena

Chapter 2 An Overview of Internet of Things in Healthcare 15
Madhvi Saxena, Subrata Dutta

Chapter 3 Universal IoT Framework 45
Joy Dutta, Sarbani Roy

Chapter 4 IoT Middleware Technology: Review and Challenges 71
Arindam Giri, Subrata Dutta, Kailash Chandra Mishra, Sarmistha Neogy

Chapter 5 IoT Service Platform 91
Prachet Bhuyan, Abhishek Ray

Chapter 6 Software Integrated Framework Design for IoT-based Applications 99
Sugyan Kumar Mishra, Anirban Sarkar

Chapter 7 Security Issues and Challenges in IoT 115
Sandeep Mahato, Kailash Chandra Mishra, Subrata Dutta, Sujoy Mistry

Chapter 8 A Framework for Delivering IoT Services with Virtual Sensors: Case Study Remote Healthcare Delivery 137
Nandini Mukherjee, Sunanda Bose, Himadri Sekhar Ray

Chapter 9 Opportunities and Challenges of IoT-Based Smart City Models toward Reducing Environmental Pollution 153
Ajanta Das, Shubham Prasad, Sameya Ashraf

Chapter 10 A Novel QoS-based Flexible Service Selection and Composition Model for Localizing the Wireless Sensor Nodes Using AHP 171

Akhilendra Pratap Singh, Arun Kumar, Om Prakash Vyas, Shirshu Varma

Chapter 11 Catalyst Is Important Everywhere: The Roles of Fog Computing in an IoT-based e-Healthcare System 195

Sankalp Nayak, Debajyoty Banik

Chapter 12 IIoT: A Survey and Review of Theoretical Concepts 223

Souptik Ghosh, Mahendra Kumar Gourisaria, Siddharth Swarup Routaray, Manjusha Pandey

Index 237

Foreword

Technological revolutions are periods in history when the dominant technologies are rapidly replaced by new solutions and processes that spur profound industrial and societal changes. These periods witness proliferation of innovation, rapid penetration of new solutions in all parts of the economy, resulting in profound changes in lifestyle and work practices. We live in an era called the Information Age, which embeds the Digital Revolution, also known as the Information and Communication (ICT) Revolution. The Digital Revolution is marked by the replacement of electromechanical and analog technologies with digital electronics. It enabled packaging of millions of logic gates into a very small physical space to build complex information processing devices. The execution of complex operations comes at a minimal infrastructure and operational costs. The ability to instantaneously transfer information between devices across the world has unfolded unprecedented capabilities, previously found only in science fiction writings and movies.

Our world is full of "smart things." Sensors, actuators, electronic brains, and communication systems are embedded in a broad spectrum of smart items and systems. Mobile devices and wearables are programmed by users to perform simple functions such as alerts or reminders, but they also perform complex processes, such as continuous monitoring of one's health. We have smart toothbrushes programmed to remind the user that it is time to brush their teeth, but they can also report on the quality and sufficiency of the oral hygiene of the user from the analysis of previous usages of the smart toothbrush. Smart technologies are scalable—multiple devices are connected into smart systems, such as smart cars, smart homes, smart offices, or smart hospital beds, to name a few. These systems are integrated into smart supersystems, such as smart buildings, smart hospitals, smart airports, or smart highways, and link them into smart megasystems, such as smart cities. The 5G connectivity enables rapid expansion of wireless sensor networks connecting multiple devices into Internet of Things (IoT) networks. A single 5G mobile phone can support at minimum 40 wide area network (WAN) bands, as well as multiple radio frequencies for wireless local area networks (WLANs). Thus, an average person already has a multiple IoT network connectivity right in his or her own pocket or handbag.

Several main challenges need to be resolved before the full benefit of the 5G and the IoT technologies can be realized. They include interoperability of devices, ultra-efficient algorithms for real-time processing of information, and the development of logic for coordinated decision-making in complex IoT networks. This book addresses the first challenge, interoperability of the IoT devices and smart systems. It covers fundamental technological issues (IoT framework, middleware, and service platform), implementation issues (software framework, security and privacy, service selection, and fog computing), and application examples (ambient-assisted living, healthcare, and environmental pollution control). It represents an informative resource for the exciting field of smart technologies.

Prof. Vladimir Brusic
Li Dak Sum Chair Professor
School of Computer Science
University of Nottingham, Ningbo, China
(China Campus)

Preface

The Internet of Things (IoT) has immense potential to transform the way we are going to live in the future, and with the advent of 5G technology, the growth of IoT is slated to increase to the order of billions of dollars. The growth is, however, not an easy one, and there were several key issues in the way of challenges. Among them, a key factor in the way of its swift development is interoperability.

IoT is now being used in such a huge variety of areas and with such diverse areas that its own diversity can be the main hindrance to its growth. Countless devices, with as many types ranging from household appliances, wearable devices, drones, and with different technical profiles and with a variety of brands of manufacturers, are expected to operate with each other to provide the services. Under such circumstances, developing the ability for the devices to seamlessly communicate with each other is a tough technical challenge and also requires a common consensus from all participating devices concerned. Therefore, interoperability in IoT is a very crucial factor for the smooth and fast development of IoT systems.

Thus the applications for IoT systems should be global for serving different industries and fields. Here, information interoperability needs to take place between enterprises, industries, regions, or countries. Interoperability in an IoT system is essential for going through layers of physical network, communication, and application functions. Different domain languages and protocols contribute to building these levels. Moreover, the heterogeneity among IoT devices arises, due to their design and development by different vendors using different systems and communication specifications. Additionally, the interoperability issue raises scalability issues in the IoT network, because connecting a new device with the network requires many configurations to operate with the existing devices. Therefore, a comprehensive approach is needed to address and solve the interoperability issues of IoT devices and services from different layers. These issues become even more important when the applications are used in critical smart systems, where failure to interact between two or more devices can lead to fatal consequences.

To understand and emphasize on the essentiality of interoperability in IoT, we consider the application of IoT in an e-Health system. In an e-Health system, doctors can access and visualize health-related data of remote patients in their mobile devices over the internet with the help of Body Area Sensor Network and medical devices attached to the patient's body. They also provide their consultation remotely over the internet, based on the patient's medical record. Here, many heterogeneous devices, such as sensors, medical devices, and networking devices, which may be provided by different manufacturers, play a role in establishing communication between the patients and the doctors. In such a scenario, interoperability among the devices is an important factor for them to work collaboratively to come up with a smart and efficient system. In case some of the devices fail to communicate with each other, it can lead to serious consequences.

This book consists of 12 chapters, which are as follows:

Chapter 1: "Internet of Things: Challenges and Its Applications" is about various issues and challenges that exist during the designing of IoT systems and the various types of applications that can benefit from implementing the IoT technology.

Chapter 2: In "An Overview of Internet of Things in Healthcare," the authors present an overview of the IoT healthcare system for collecting, monitoring, integrating, and interoperating data wirelessly in medical services. The core involvement of this work is to reduce the physical presence of patients and provide deterministic results and analysis of traditional healthcare and its challenges and the advantages of IoT healthcare over traditional approaches.

Chapter 3: "Universal IOT Framework" is a review of various existing unified IoT frameworks for various applications and the requirements and modifications needed at the various layers to implement them and the associated challenges.

Chapter 4: "IoT Middleware Technology: Review and Challenges" reviews the current middleware technology in IoT and discusses the challenges associated with the development of a suitable middleware technology, with respect to a common IoT framework.

Chapter 5: "IoT Service Platform" focuses on various aspects of IoT service platforms in the sections: introduction to IoT platform, referential IoT platforms, challenges of the IoT platform, and summary and future of the IoT platform.

Chapter 6: "Software Integrated Framework Design for IoT-Based Applications" deals with an emerging area of research in IoT, which is software architecture (SA). The objective of SA is to design a framework that meets the requirements set for certain quality attributes, but there is an absence of benchmarks for IoT structures, especially with regard to modern IoT frameworks. It also deals with issues such as developing an appropriate IoT framework for the description of services, and additionally the integration of the IoT foundation with BPM and a good evaluation procedure of software architecture with quality attributes.

Chapter 7: "Security Issues and Challenges in IoT" is based on recent IoT security research and provides a basis for better understanding the future potential, deals with key challenges for the construction of IoT, security issues, and vulnerability, causing attacks and threats.

Chapter 8: "A Framework for Delivering IoT Services with Virtual Sensors: Case Study of Remote Healthcare Delivery" is a case study of a remote healthcare delivery application, where a kiosk-based application is discussed and its implementation and its impact on the society. Although sensing and cloud technologies are two primary ingredients for developing IoT services, there are several challenges that need to be handled for delivering uninterrupted sensing services over the internet. They include handling heterogeneous sensing devices, non-interoperability issues due to different standards and protocols, acquiring sensing devices from external network by IoT applications, reducing data transmission from the sensing devices to cloud, and handling huge amounts of unstructured and semi-structured data in cloud.

Chapter 9: "Opportunities and Challenges of IoT-Based Smart City Models toward Reducing Environmental Pollution" concentrates on continuous monitoring of Air Quality Index (AQI) for various regions of Kolkata, India. The analysis of AQI data for the last six months is presented to highlight any anomalies between the data, along with the maximum of lowest AQI and minimum of highest AQI to alert citizens for the specific regions. It is the proposed action plan for reducing air pollution to achieve Smart City Mission in Kolkata, India.

Chapter 10: "A Novel QoS-Based Flexible Service Selection and Composition Model for Localizing the Wireless Sensor Nodes Using AHP" proposes a novel QoS-based flexible service selection and composition model that provides a platform for users to select the best possible service. For the design goal, network and application requirements are illustrated, and a mathematical formulation is presented using analytical hierarchy process (AHP). The proposed framework is validated and tested, based on given QoS parameters, and the results demonstrate the efficacy of the proposed method.

Chapter 11: "Catalyst Is Important Everywhere: The Roles of Fog Computing in IoT-Based e-Healthcare" the fog computing notion in the IoT healthcare system brings cloud layers closer to IoT medical devices. The concept is to introduce a geographically distributed mediator layer of intelligence connecting sensor nodes and clouds, which can manage the medical device data efficiently, and provide a faster response. It presents the integration of fog computing in the IoT healthcare system, highlighting its strengths and implementation challenges. We then finally draw attention toward an IoT-based early warning score (EWS) health monitoring system case study.

Chapter 12: "IIOT: A Survey and Review of Theoretical Concepts" introduces the concept of IIoT. IIoT refers to interconnected sensing devices, machines, and other devices connected together with a computer for industrial usage, including production and energy management. The vast set of scope for IoT devices is basically segregated as consumer purposes, commercial applications, industrial usage, and infrastructure space implementation, with smart systems, elderly care, medicine and health resources, transportation, communication, and automation. Before writing this book, we thoroughly studied the existing state-of-the-art scenario in this area. We found very few books have completely discussed interoperability issues in IoT systems and their possible solutions. No book has talked about scalability and security issues as a consequence of interoperability issues in IoT systems and proposed solutions for them. In addition to this, we decided to consider how to improve the performance of an IoT system. So we have proposed the concept of fog computing and its possible use in an IoT system. We also thought that we should end the book by discussing some real-life IoT systems. So we presented an e-Health system as a possible real-life IoT system.

In this book, we have organized the chapters to discuss challenges and issues in interoperability among the IoT devices and their solutions. We have also dedicated a few chapters, which discuss the scalability and security issues in an IoT network and provide solutions for the interoperability of new devices within the existing IoT system. The book also discusses the possible techniques for the use of interoperable IoT network in different areas. We have also presented an e-Health system and a smart city application, as case studies to help the readers get a better understanding of the solutions of interoperability and scalability of an IoT system, as well as their uses to make a system smarter. At the end of the book, we have discussed fog computing as a viable approach for improving the performance of an IoT system and its application in the e-Health system.

We express our sincere gratitude to all the authors for their valuable contributions in compiling this book. Without their research efforts and documentations, it would not have been possible to complete the book. We also thank our reviewers for their great efforts taken. We hope this book will be very helpful for academics as well as industry persons who wish to continue their work in advancements of IoT in smart applications. We thank Mr. Rahul Acharya, Senior Manager, Ericsson Global Services India Pvt. Ltd. for the excellent cover design. We also thank our publishers, Taylor and Francis, for bringing our effort in the light of the day that it deserves. Their staff were very friendly and helpful with their insightful suggestions throughout our journey through the last one year during the creation of this work.

Monideepa Roy
Pushpendu Kar
Sujoy Datta

About the Authors

Dr. Monideepa Roy obtained her master's degree in mathematics from IIT Kharagpur and her PhD in CSE from Jadavpur University. Currently she is working as Associate Professor at KIIT Deemed University, Bhubaneswar, since the last seven years. Her areas of interest include wireless networks, mobile computing and WSNs, and cognitive WSNs, machine learning, remote sensing, artificial neural networks, and combinatorics. At present, she has five research scholars working with her in these areas. She has several publications in conferences and journals. She has been the Organizing Chair of the first two editions of the International Conference on Computational Intelligence and Networks CINE 2015 and 2016, and the International Conference on Mathematics and Computing (ICMC 2019). She has also organized several workshops and seminars. She has successfully completed an MeiT-sponsored five-year project on Remote Healthcare as the Principal Investigator.

Dr. Pushpendu Kar is Assistant Professor in the School of Computer Science at the University of Nottingham, Ningbo, China (China Campus of the University of Nottingham, UK). Prior to this, he was Research Fellow in the Department of ICT and Natural Sciences at Norwegian University of Science and Technology (NTNU), Norway, in the Department of Electrical and Computer Engineering at National University of Singapore (NUS), and in the Energy Research Institute at Nanyang Technological University (NTU), Singapore. Dr. Kar has completed all his PhD, Master of Engineering, and Bachelor of Technology in Computer Science and Engineering. He has also completed the Sun Certified Java Programmer (SCJP) 5.0, one professional course on Hardware and Networking, two professional courses on JAVA-J2EE, Finishing School Program from National Institute of Technology, Durgapur, India, and UGC-sponsored refreshers course from Jadavpur University, India. Dr. Kar went for a research visit to Inria in Paris, France. Dr. Kar was awarded the prestigious Erasmus Mundus Postdoctoral Fellowship of European Commission, ERCIM Alain Bensoussan Fellowship of European Union, and SERB OPD Fellowship of Department of Science and Technology, Government of India. He has received many travel grants to attend conferences and doctoral colloquiums. Dr. Kar has more than nine years of teaching, research, experiences, including in a couple of highly reputed organizations around the world. He has worked as a software professional in IBM for one and one-half years. Dr. Kar is the author of more than 30 scholarly research papers, which he has published in reputed journals, including ACM TAAS, IEEE TNSM, IEEE Systems Journal, IEEE Sensors Journal, Journal of Building and Environment, conferences including ICC, TENCON, IECON, PEDS, and IT magazines. He is also the inventor of four patents. He has participated in the program committees of several conferences, worked as a team member to organize short-term courses, and delivered a few invited talks. He is a regular reviewer of IEEE, Elsevier, Wiley, and Springer journals and conferences.

Mr. Sujoy Datta has done B Tech from Burdwan University. He has obtained his MSc and M Tech. from IIT Kharagpur in 2010 and 2012, respectively. Currently he is working as Assistant Professor in the School of Computer Engineering, KIIT Deemed University, since the last seven years. His areas of research include probability, wireless networks, security, threshold cryptography and neural networks, and PAC learning. He has several publications in various conferences and journals. He has co-organized several workshops and international conferences, CINE 2015, and ICMC 2019.

1 Internet of Things
Challenges and Its Applications

Zeenat Rehena
Aliah University, West Bengal, India

CONTENTS

1.1 The Internet of Things ... 1
1.2 IoT Components and Technologies ... 2
 1.2.1 Components ... 3
 1.2.2 Technologies ... 4
1.3 IoT Architecture ... 5
1.4 IoT Analytics Life Cycle ... 6
1.5 Challenges of IoT ... 7
1.6 Applications of IoT ... 8
1.7 Future Research Direction ... 11
1.8 Conclusion ... 12
References ... 13

1.1 THE INTERNET OF THINGS

In the last few decades, improvements in technology have been made tremendously, especially in Information and Communication Technology (ICT). With this advancement in internet, different emerging technologies, namely, distributed computing, grid computing, cloud computing, and ubiquitous computing have made significant impact into the computing world [1].

The Internet of Things (IoT) is another astonishing paradigm toward modern technology and the computing world. It is an emerging technology with applications in many different fields, which is bringing the total world into its own domain.

Nowadays, IoT has become the most common and popular trend in marketing. There are several definitions of IoT in the literature, proposed by different researchers, academicians, and organizations in their own way. International Telecommunication Union-Global Standards Initiative (ITU-GSI) [2] has defined IoT as: "the network of physical objects—devices, vehicles, buildings and other items—embedded with electronics, software, sensors, and network connectivity that enables these objects to collect and exchange data."

Therefore, it connects all the objects or things with the underlying technology and makes interaction with each other with the help of the internet. Connecting objects or things might be wireless, radio frequency identification (RFID), or sensors, actuators,

TABLE 1.1
Population versus Connected Devices with Ratio of Connected Devices per Person

Year	World Population (in Billion)	Connected Devices (in Billion)	Ratio of Connected Device per Person
2003	6.3	0.5	0.08
2010	6.8	12.5	1.84
2015	7.2	25	3.47
2020	7.6	50	6.58

and mobile phones. These technologies are used for identification of items and sensing of the environment. So, IoT links various things, using electronic sensors, turning them into a smart thing by allowing them to see, hear, think, and compute tasks, etc. Its major goal is to transform things into smart things, objects into smart objects with the help of embedded devices, wireless sensor network (WSN), pervasive computing, ICT, and applications, and it is concerned in every aspect of the real world [3].

As per the Cisco Internet Business Solutions Group (IBSG) [4], "IoT is simply the point in time when more things or objects were connected to the Internet than people." It is observed that in 2003, the world population was counted at 6.3 billion, whereas connected devices were only 0.5 billion. The world population was 6.8 billion, whereas connected devices were 12.5 in the year 2010. Similarly, in 2015, the world population was 7.2 billion and the connected devices count had become tripled to the population count, i.e., 25 billion. In 2020, the population has become 7.6 billion, and connected devices have reached 50 billion [4]. The world population versus connected devices with ratio of connected devices per person is shown in Table 1.1.

IoT is matured now and it has explored its wings outside the laboratory. It has great impact in plenty of application areas, which also include environment monitoring, manufacturing, healthcare, smart cities, smart living, and many more [1]. The general diagram of IoT is shown in Figure 1.1.

The rest of this chapter is organized as follows: Basic components and underlying technologies are discussed in Section 1.2. In Section 1.3, the generic architecture of IoT is described. The IoT analytics life cycle is briefly discussed in Section 1.4. Major challenges of IoT are discussed in Section 1.5. The challenges faced from different applications are explained in this section. In Section 1.6, different applications of IoT are elaborated and discussed. Future directions of research in IoT and concluding remarks are given in Sections 1.7 and 1.8, respectively.

1.2 IoT COMPONENTS AND TECHNOLOGIES

In this section, IoT components and enabling technologies [5, 9] for the underlying components are discussed briefly. An IoT system consists of some functional units to facilitate various functions to the system, such as sensing, identification, actuation, communication, and management, while enabling technologies are those which will activate the aforementioned components and composite into a single system.

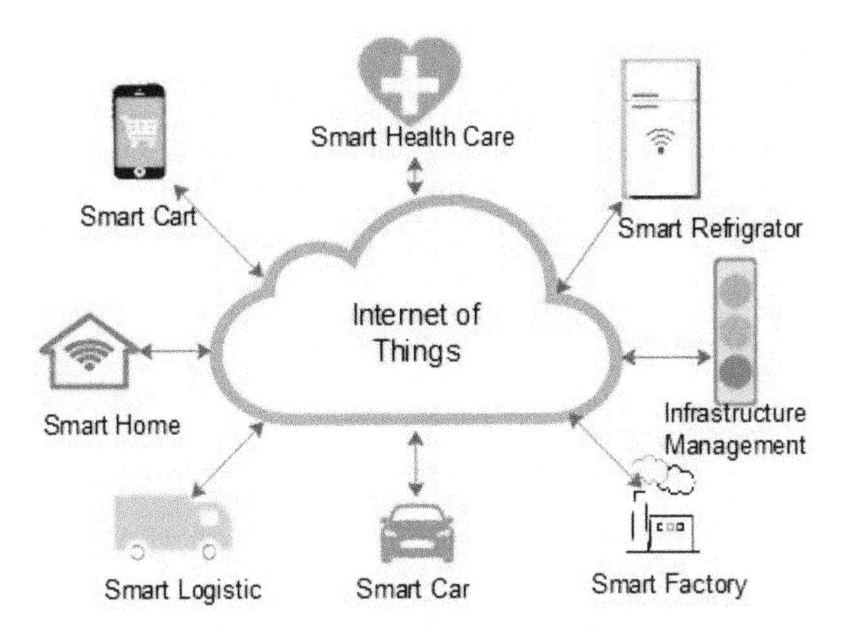

FIGURE 1.1 General representation of the Internet of Things.

1.2.1 Components

The IoT components are described below [5]:

- **IoT device:** IoT devices are any objects or things that have sensors attached and make them able to sense, control, and monitor activities. Then the devices exchange data with other connected devices in the system, and data processing is done within the processing components. There are several types of IoT devices, such as wearable sensors, actuators, software, smart watches, LED lights, automobiles, and industrial machines. For example, the IOT system of a car identifies the traffic on the road and automatically informs the person that there is an impending delay during the journey.
- **Communication infrastructure:** The communication unit does the communication between IoT devices and remote processing units or cloud server. IoT communication protocols generally work in the data link layer, network layer, transport layer, and application layer.
- **Services:** An IoT system offers several types of services for IoT devices, such as device modeling, device control, data processing, etc.
- **Security and management:** Security components secure the IoT system by invoking several functions, such as authentication, authorization, privacy, message integrity, content integrity, and data security. On the other hand, the management unit controls and manages the IoT system.
- **Application:** The application module acts as an interface between users and the IoT system. It enables users to visualize and analyze the present scenario system status and occasionally predicts future prospects.

1.2.2 Technologies

- **Radio-frequency identification:** RFID is used to identify objects wirelessly without line of sight. RFID consists of a reader and one or more tags. The RFID reader is used to read that data. It employs two-way radio transmitter-receivers to identify and track tags associated with objects. There are two types of RFID tags, active and passive. Active tags have a power source and passive tags do not have any power source. Passive tags draw power from the reader's interrogation signal during the communication to the RFID reader, and thus it is cheap and has a long lifetime [6, 7]. It has many applications, especially in retail and supply chain management, in transportation for buying or replacement of tickets, registration stickers, etc.

  The RFID technologies are also having two different variants, such as near and far [6]. A near-RFID reader uses a coil through which alternating current is passed to generate a magnetic field. On the other hand, far-RFID uses a dipole antenna in the reader, which propagates electromagnetic (EM) waves. RFID technology is used in supply chain management, access control, identity authentication, and object tracking, etc.
- **Wireless sensor networks:** WSN is the boon in low-power integrated circuits and wireless communications. The most important component of WSN is sensors. These are low-cost, battery-powered tiny devices, which make WSNs in remote sensing applications. A WSN consisting of thousands of such intelligent sensors is able to sense, process, and disseminate the valuable data of the monitoring environment [8]. Sensor nodes are gathered from the environment and shared among other sensor nodes and then sent to a distributed or centralized system via sink node for analytics. The applications like environment monitoring, habitat monitoring, structural health monitoring, and health monitoring are most popular in WSN.
- **Addressing schemes:** Uniform Resource Name (URN) is mostly used in IoT. Several replicas of resources are generated by the URN that is accessed through the URL [9]. Different types of sensor nodes which are communicated to the central node (sink) are made addressable by URN. On the other hand, IPv6 also gives a very good solution to accessing the resources uniquely and remotely. The standard is independent of the underlying physical layer and frequency band. It can have utilization over different communications platforms, including Ethernet, 802.15.4, WiFi, and subl GHz ISM (Industrial, Scientific, and Medical) radio channels. IPv6 is useful, especially for developing building and home automation. It provides the fundamental transport scheme to create complex control systems and to connect with devices cost-effectively through a low-energy wireless network.
- **Data storage and analytics:** Before going to the application layer, the data have to be stored and processed efficiently. There are basically two

kinds of software components, middleware and applications. The processing or middleware does all the computing steps to create a meaningful context. Artificial intelligence (AI), machine learning algorithms, neural networks, and fusion algorithms are necessary to develop the decision-making system.

1.3 IoT ARCHITECTURE

There is no such single standard on architecture for IoT, which is agreed universally. Different architectures have been proposed by different researchers across the world. In this section, a five-layer architecture is described briefly. This five-layer architecture defines the main idea of the IoT, and it is shown in Figure 1.2. It has five layers, namely, perception layer, network layer, processing of middleware layer, application layers, and business layer [1, 10].

a. The *perception layer* is the physical layer where sensors are attached with the objects for sensing and collecting information about the environment. They sense some ambient conditions of the environment and complete the tasks, collaborating with other smart objects in the environment. The sensed information can be temperature, humidity, orientation, location about the objects, motion, vibration, acceleration, chemical changes in the air, etc.

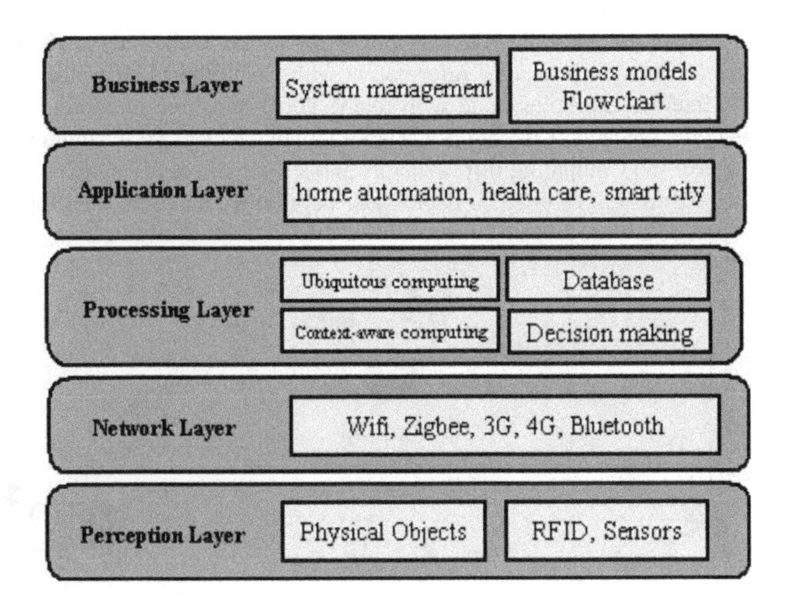

FIGURE 1.2 Generic architecture of IoT (five-layer).

b. The next layer is the *network layer* and is responsible for connecting to other smart things, network devices, and servers. That means that it transfers the data from the perception layer to the processing layer. The transfer of data can be done through wired or wireless network and the underlying technology could be 3G, UMTS (Universal Mobile Telecommunications Service), WiFi, Bluetooth, infrared, ZigBee, etc., depending upon the sensor devices.
c. The next higher layer is the *processing layer,* which is also known as the *middleware* layer. It stores, analyzes, and processes huge amounts of data that come from the lower layer, i.e., network layer. It utilizes many technologies, such as databases, context-aware computing, cloud computing, and Big Data-processing modules. The incoming data are heterogamous and have different data formats and size. Initially, it invokes context-aware computing analytics to generate a meaningful single context with those huge data. And then the decision-making unit makes a decision based upon these contexts.
d. The next upper layer is the *application layer,* and it is responsible for delivering application-specific services to the user. It defines various applications in which the IoT can be deployed, for example, smart homes, smart cities, smart health, intelligent agriculture, smart car, and smart environment, etc.
e. The final and top layer is the *business layer.* It is responsible for the management of the total IoT system. It includes services, system management, business models, and services to the applications.

1.4 IoT ANALYTICS LIFE CYCLE

IoT is a platform for embedded devices and is connected to the internet. Then the devices can collect and exchange information with each other. Hence, IoT enables devices to interact and collaborate with them. There are three main components in IoT analytics, namely, Collection of Data (Things or Objects), Communication Infrastructure, and Computing Infrastructure. This life cycle of IoT analytics is represented in Figure 1.3. In order to deploy and develop IoT analytics, the following life cycle phases are used.

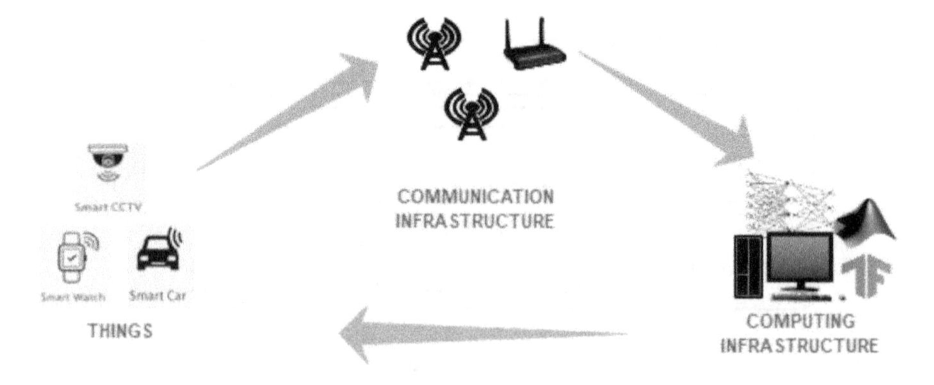

FIGURE 1.3 Life cycle of IoT.

a. **Collection unit:** This unit comprises smart devices, sensors, product systems, and other objects or things. It collects the raw data from these devices. Before sending data, it needs to be validated for data integrity, accuracy, and consistency of the sensed data. Therefore, this component addresses several challenges of IoT.
b. **Communication unit:** The internet is the backbone of the communication infrastructure. It connects collection infrastructure to a computing unit.
c. **Computing unit**: This unit processes and analyzes all the raw data and generates a new form of information. It ensures the IoT data or knowledge according to the needs of the application. The processed data are delivered to the smart devices in real time.

1.5 CHALLENGES OF IoT

This section explains the main challenges associated with the development and deployment of IoT.

As IoT consists of any things or any objects and these objects are connected through the communication systems, so that it generates enormous data all the time. It has deep interconnection with the Big Data. IoT has huge benefits, but it also faces many key challenges [1, 10, 21]. These are briefly described below.

- **Heterogeneity of IoT data:** IoT applications are ubiquitous in nature. Therefore, they produce a deluge of data, which are heterogeneous in terms of their formats and semantics. Hence, IoT analytics applications might depend on these types of data. Big Data technologies provide the means for dealing with this heterogeneous data. However, accessing and managing IoT data sources (including sensors and other types of internet connected devices) is a big challenge.
- **Varying data quality**: Since data are heterogeneous, several IoT streams are noisy and incomplete, and that produces uncertainty in applications. Therefore, some mechanisms are needed in order to maintain data inputs, especially in cases of unreliable sensors.
- **Real-time nature of data**: Several applications may be processed almost in real time, thus IoT needs high-velocity data. It is very important to focus on changes in data patterns rather than dealing with all the sensed data that have been sent from a given sensor.
- **Identifying objects:** Each object or sensor has a unique identification number (Id) over the internet. To connect billions of objects or sensors for gathering information, it is very important to know their unique IDs. Therefore, it is required to have an efficient naming and identity management system. This enables it to dynamically assign and manage the unique identification numbers when it encounters many numbers of objects.
- **Privacy and security**: Barcodes and radio frequency identifier are used as an object identifier in IoT [11]. Every object will carry these identification tags, and it is necessary to take proper privacy preservation techniques and prevent unauthorized access. Further, data encryption mechanism is

needed when sensed data are transferred from sensor to the data processing or computing unit for integrity of the information.

- **Greening of IoT:** Sensor nodes transmit data through a wireless medium. On the other hand, due to a limited available spectrum, billions of sensors need efficient dynamic spectrum allocation techniques for communication over a wireless medium. Moreover, network energy consumption increases day by day with the increase in high data rates. This causes a considerable amount of network energy consumption. Thus, green technologies require to be adopted for network devices to become as energy efficient as possible.

1.6 APPLICATIONS OF IoT

IoT has potential impact in almost every aspect of our daily life. It has lots of applications in different diverse areas [5]. The popular and common application areas, such as environment monitoring, home automation, medical support system, and smart cities, are summarized as below. Figure 1.4 shows the tree representation of IoT application areas.

1. **Environment monitoring:** This domain includes several applications, such as prediction of natural disasters, water scarcity monitoring, agriculture, etc. Environmental parameters such as temperature, humidity, and air

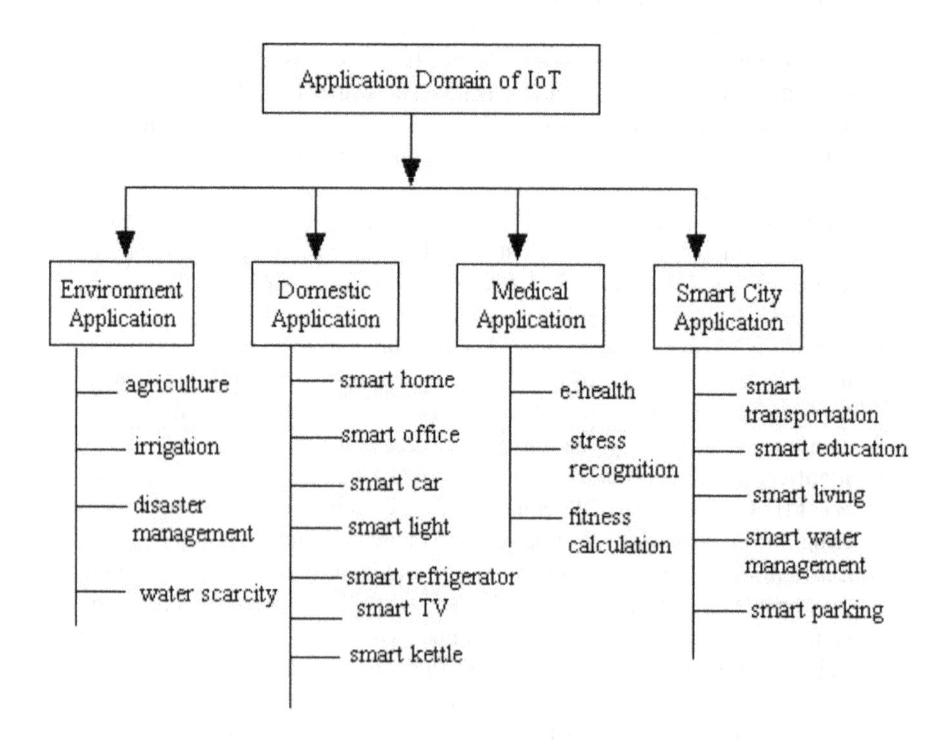

FIGURE 1.4 Tree representation of IoT application areas.

pressure are sensed by the sensors, and data can be used for the applications. IoT-based intelligent agriculture monitoring [12] and automated irrigation according to weather conditions [1, 10] uses sensors data from the fields and helps in the efficient production of crops. Environmental parameters, such as temperature, soil information, and humidity, are measured in real time and sent to a server for analysis. Sensed values are fed into the microcontroller, and depending upon the threshold values of different parameters of the monitoring field, it acts accordingly. Then the farmer is informed through mobile phone text messages whether the field needs particular attention or not. Hence, the results improve the crop quality. Another application, like production using greenhouses [13], is an example of IoT in agriculture. Reduction of pesticide usage in agriculture is also detected using IoT applications [14].

Further, disaster management applications will help to predict the occurrence of natural calamity beforehand and to take necessary actions in advance. Again, IoT can help to detect the water scarcity at different places. WSNs have been used for such type of applications.

2. **Domestic applications:** Smart home, smart office, smart refrigerator, and smart cars are examples of domestic applications of IoT. Today, smart homes are very popular, especially for the elderly and differently abled people, due to its concern toward improvement of quality of life of ours and security of our homes. In smart home applications, various sensors are deployed and connected with several objects to provide intelligent services to the user [1]. There are several applications which help by turning off lights and TV, AC and other electronic gadgets automatically. Hence, it saves electricity. Energy saving in smart homes [15] can be achieved using sensors and context-aware computing. The sensors collect data from the environment (light, temperature, humidity, gas, and fire events), and then the sensed data are fed to the context-aware computing analytics. The sensed data are heterogeneous in nature since there are different types of sensors used. Context-aware computing processes these data and generates a meaningful context, and based on this context, the service will be invoked.
3. **Medical applications:** IoT has applications in the medical sector also. It helps to save lives and improve the quality of life. These applications monitor health parameters, monitoring patients' activities, providing support systems for independent living, monitoring medicine intake, etc. Nowadays, there are many wearable devices available in the market, which monitors the patient's health parameters. Figure 1.5 represents an IoT-based health monitoring application system. Health applications are also very helpful for elderly people who are living independently at home. Wearable devices continuously monitor and sense the health parameters, and if any abnormal condition arises, the alarm is generated. Other popular medical applications of IoT are stress recognition [16, 17] and fitness recognition [18], which measures stress level and amount of fitness of users, respectively. The Smartphone's sensor is used for it. It captures the whole-day information of a person, such as visits during the day, amount of physical activities

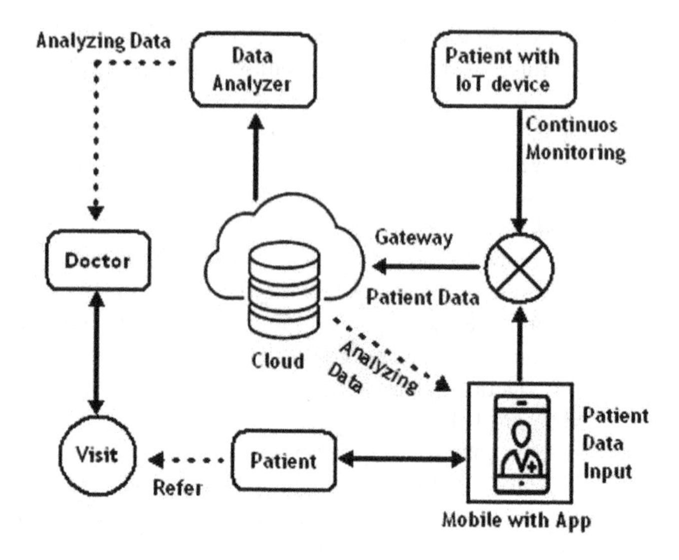

FIGURE 1.5 IoT-based health monitoring application.

done, duration of sleep and rest, and interaction with other people. With the help of the interaction with the person and intelligent algorithms, it analyses the amount of stress of that person.

4. **Smart city applications:** The smart city applications are citizen-centric, where citizens are the main users of it. The six key components of smart city are: (i) smart people, (ii) smart city economy, (iii) smart mobility, (iv) smart environment, (v) smart living, and (vi) smart governance [19]. These six building blocks are closely interlinked and contribute to the smart city system. Among these, some are discussed here briefly.
   a. **Smart transportation:** Smart transport applications are used to manage daily traffic in cities using sensors and intelligent information processing systems. The main aim of intelligent transportation systems (ITSs) [20] is to minimize traffic congestion on roads, so that it reduces journey time, ensures hassle-free parking, and updates notifications about road accidents or congestion to the driver or citizen. Several types of sensors are used to accomplish these tasks, such as GPS sensors for location, accelerometers for speed, gyroscopes for direction, RFIDs for vehicle identification, infrared sensors for counting passengers and vehicles, and cameras for recording vehicle movement and traffic.
      i. **Traffic surveillance and management:** For traffic surveillance, IoT devices such as GPS sensors, RFID devices, and cameras are attached to vehicles and the vehicles are connected by a network to each other. This can help to detect vehicles and can analyze traffic patterns so that future traffic conditions can be predicted. Several IoT-based mechanisms like Traffic Congestion Control Monitoring

[21] system and Intelligent Parking Management [22] also have a great impact on traffic congestion and thus reduces pollution in the air. It helps drivers or users for hassle-free parking while going to visit hospitals, shopping malls, offices, or restaurants. In smart traffic lights management, traffic lights equipped with sensing, processing, and communication capabilities are called smart traffic lights. These lights sense the traffic congestion at the intersection and the amount of traffic going each way. This information can be analyzed and then sent to neighboring traffic lights or a central controller. It is possible to use this information creatively. For example, in an emergency situation, the traffic lights can preferentially give way to an ambulance or fire-bridged or police van.

b. **Smart water management:** It is very important to manage water resources efficiently because of water scarcity in most parts of the world. Currently, India is in an alarming situation. Therefore, most cities under the smart city mission of the Government of India are opting for smart solutions that place a lot of meters on water supply lines. Although smart metering is implemented and using in several cities throughout the world, still it is in a nascent stage in India. Smart water metering systems are also used in conjunction with data from weather satellites and river water. They can also help us predict flooding.
c. **Smart living:** Smart living combines *s*ocial life and entertainment in a person's life. Many applications have been developed, which keep track of such human activities. It records and celebrates local history, culture, and nature. It should be a vibrant downtown, 24 hours and 7 days a week. Smart living also ensures natural and cultural assets to build a good quality of life.

Sections 1.5 and 1.6 constitute the several challenges and application areas of IoT. Based on the above discussions, the application domains of IoT are summarized in Table 1.2.

1.7 FUTURE RESEARCH DIRECTION

Currently, there are so many developments of applications where IOT is used dominantly [23]. The applications includes risk management processes, improvement of data transfer security systems on the internet and support of business management processes, crisis management in urban agglomerations, smart city, cargo management, etc. The IoT technology will gradually expand in the situation of creating system solutions that combine several ICT technologies. Still, there are a lot of challenges in various applications that need to be solved in the near future. These include:

1. How AI and machine learning techniques along with IoT are combined to make Artificial Intelligent Internet of Things (AI-IoT).
2. How Big Data analytics, edge computing, and IoT are combined and make solutions for different applications, specifically for healthcare solutions.

TABLE 1.2
IoT Application Domains

Applications	Network Size	Users	Energy	Internet Connectivity	IoT Devices	Data Management
Smart home/ office	Small	Few, family members	Rechargeable battery	WiFi, 3G, 4G LTE backbone	RFID, WSN	Local server
Smart city	Medium/ large	Citizens, govt., policy makers	Rechargeable battery, energy harvesting techniques	WiFi, 3G, 4G LTE backbone	RFID, WSN	Shared server/ cloud server
Smart agriculture	Medium/ large	Small, farmers, landowners	Energy harvesting techniques	WiFi, satellite communication	WSN	Local server, shared server
Smart transpor-tation	Large	Citizens, govt., policy makers/ traffic police	Rechargeable battery, energy harvesting techniques	WiFi, satellite communication	RFID, WSN,	Shared server
Smart water	Medium/ large	Citizens, govt.	Energy harvesting techniques	Satellite communication	Sensors	Shared server
Smart health	Medium	Patients, doctors, caregiver	Rechargeable battery, energy harvesting techniques	WiFi, 3G, 4G LTE backbone	RFID, BAN, wearable sensors	Local server, shared server/cloud server

3. AI, edge computing, block chain can be combined with IoT for meeting the future demand.
4. AI and machine learning algorithms are used for security and privacy of the systems, and researchers have already started working in this direction. Still, there are possibilities to explore in that direction.

1.8 CONCLUSION

The IoT is an emerging technology of internet, which connects every smart object with every other object. This has added a new potential into the internet, thus making a smarter and more intelligent planet.

In this chapter, firstly the background and definition of IoT are given. Secondly, thorough discussions on IoT components and architecture are discussed. Next, it demonstrates IoT analytics and addresses some key research challenges associated with the IoT technology. Finally, it discussed possible future applications and aiming at the future directions of IoT applications domains.

REFERENCES

1. P. Matta, and B. Pant, "Internet-of-things: genesis, challenges and applications," Journal of Engineering Science and Technology, vol. 14, no. 3, pp. 1717–1750, 2019.
2. International Telecommunication Union (ITU). (2015). Internet of things global standards initiative. Retrieved on January 8, 2017, from http://www.itu.int/en/ITU-T/gsi/iot/Pages/default.aspx.
3. E. Umamaheswari, and D. Ajay, "Scope of internet of things: a survey," Asian Journal of Pharmaceutical and Clinical Research, vol. 10, no. 13, pp. 187–90, 2017. https://doi.org/10.22159/ajpcr.2017.v10s1.19633.
4. D. Evans, The Internet of Things: How the Next Evolution of the Internet Is Changing Everything. San Jose, CA: Cisco, 2011.
5. P. Sethi, and S. R. Sarangi, "Internet of things: architectures, protocols, and applications," Journal of Electrical and Computer Engineering, vol. 2017, pp. 1–23, 2017.
6. R. Want, "An introduction to RFID technology," IEEE Pervasive Computing, vol. 5, no. 1, pp. 25–33, 2006.
7. X. Zhu, S. K. Mukhopadhyay, and H. Kurata, "A review of RFID technology and its managerial applications in different industries," Journal of Engineering and Technology Management, vol. 29, no. 1, pp. 152–167, 2012.
8. Z. Rehena, K. Kumar, S. Roy, and N. Mukherjee, Applications of Wireless Sensor Network in Forest fire detection, *2nd India Disaster Management Conference*, New Delhi, 2009.
9. J. Gubbi, R. Buyya, S. Marusic, and M. Palaniswami, "Internet of things (IoT): a vision, architectural elements, and future directions," Future Generation Computer Systems, vol. 29, no. 7, pp. 1645–1660, 2013. https://doi.org/10.1016/j.future.2013.01.010.
10. R. Khan, S. U. Khan, R. Zaheer, and S. Khan, "Future Internet: The Internet of Things Architecture, Possible Applications and Key Challenges," *2012 10th International Conference on Frontiers of Information Technology*, Islamabad, 2012, pp. 257–260.
11. Eu-china Joint Whitepaper on Internet of Things Identification, November 2014.
12. Md Ashifuddin Mondal, and Zeenat Rehena, "Iot Based Intelligent Agriculture Field Monitoring System," *8th International Conference on Cloud Computing, Data Science & Engineering (Confluence)*, pp. 625–629, IEEE, 2018.
13. J.-C. Zhao, J.-F. Zhang, Y. Feng, and J.-X. Guo, "The study and application of the IOT technology in agriculture," in Proceedings of the 3rd IEEE International Conference on Computer Science and Information Technology (ICCSIT '10), pp. 462–465, Chengdu, China, July 2010.
14. G. Zhao, Y. Guo, X. Sun, and X. Wang, "A system for pesticide residues detection and agricultural products traceability based on acetyl cholinesterase biosensor and internet of things," International Journal of Electrochemical Science, vol. 10, no. 4, pp. 3387–3399, 2015.
15. D. M. Han, and J. H. Lim, "Design and implementation of smart home energy management systems based on ZigBee," IEEE Transactions on Consumer Electronics, vol. 56, no. 3, pp. 1417–1425, 2010.
16. R. Wang, F. Chen, Z. Chen et al., "Student life: assessing mental health, academic performance and behavioral trends of college students using smartphones," in Proceedings of the ACM International Joint Conference on Pervasive and Ubiquitous Computing (UbiComp '14), pp. 3–14, Seattle, Wash, September 2014.

17. K. Frank, P. Robertson, M. Gross, and K. Wiesner, "Sensor-based identification of human stress levels," in Proceedings of the IEEE International Conference on Pervasive Computing and Communications Workshops (PerCom Workshops '13), pp. 127–132, San Diego, CA, March 2013.
18. M. Sundholm, J. Cheng, B. Zhou, A. Sethi, and P. Lukowicz, "Smart-mat: recognizing and counting gym exercises with low-cost resistive pressure sensing matrix," in Proceedings of the ACM International Joint Conference on Pervasive and Ubiquitous Computing (UbiComp '14), pp. 373–382, Seattle, WA, September 2014.
19. L. Anthopoulos, M. Janssen, and V. Weerakkody, "A Unified Smart City Model (USCM) for smart city conceptualization and benchmarking," International Journal of Electronic Government Research (IJEGR), vol. 12, no. 2, pp. 77–93, 2016.
20. Zeenat Rehena, and Marijn Janssen. "Towards a framework for context-aware intelligent traffic management system in smart cities," in Companion Proceedings of the Web Conference 2018 (WWW '18). International World Wide Web Conferences Steering Committee, Republic and Canton of Geneva, CHE, pp. 893–898, 2018. https://doi.org/10.1145/3184558.3191514.
21. Md Ashifuddin, and Zeenat Rehena, "An IoT-based congestion control framework for Intelligent Traffic Management System," in Advances in Intelligent System and Computing, Springer, 2019.
22. Zeenat Rehena, Md Ashifuddin Mondal, and Marijn Janssen. "A multiple-criteria algorithm for smart parking: making fair and preferred parking reservations in smart cities," in Proceedings of the 19th Annual International Conference on Digital Government Research: Governance in the Data Age (dg.o '18). Association for Computing Machinery, New York, NY, Article 40, pp. 1–9, 2018. https://doi.org/10.1145/3209281.3209318.
23. P.P. Ray, "A survey on Internet of Things architectures," Journal of King Saud University—Computer and Information Sciences, vol. 30, no. 3, pp. 291–319, 2018. https://doi.org/10.1016/j.jksuci.2016.10.003.

2 An Overview of Internet of Things in Healthcare

Madhvi Saxena and Subrata Dutta
NIT Jamshedpur, Jharkhand, India

CONTENTS

2.1 Introduction 16
 2.1.1 Internet of Things 17
 2.1.2 Wireless Sensor Network 17
 2.1.3 Body Area Sensor Network 17
 2.1.4 Smart Healthcare 18
2.2 Related Work 20
2.3 Overview of Internet of Things 23
 2.3.1 Architecture of IoT 23
 2.3.2 IoT Key Features 24
 2.3.3 IoT Advantages 24
 2.3.4 IoT Disadvantages 25
 2.3.5 Components and Connectivity of IoT 25
 2.3.6 Sensors and Actuation 26
 2.3.7 Applications of IoT 27
 2.3.8 Comparison Between Traditional and IoT Network 27
2.4 Internet of Things In Healthcare 28
 2.4.1 Model for IoT Healthcare Systems 29
 2.4.2 Wearable Devices and Medical Sensor 30
 2.4.3 Communication Between Devices 32
  2.4.3.1 Short-Range Communication 32
  2.4.3.2 Long-Range Communication 33
 2.4.4 Radio-Frequency Identification 35
 2.4.5 Cloud in Healthcare 36
 2.4.6 Big Data Management 37
 2.4.7 Security 38
2.5 Different Case Studies 38
2.6 Future Scope 40
2.7 Conclusion 40
References 41

2.1 INTRODUCTION

Human life in today's world is totally driven by the internet and their applications. The broad range application of the internet includes the method by which most of the electronic devices can connect to the remote location on the internet and could be used from the remote location. The basic concept of IoT (Internet of Things) includes the various types of sensors, which have the capability to record the data using the networks. These networks help in the recording of the obtained data and the controlling of the connected devices. In this era of technology, IoT is highly demanding in the field of healthcare, manufacturing, and various industrial purposes [1]. If we think about the overall functioning of any device with IoT infrastructure or without IoT infrastructure, then there is no basic difference. The main purpose of the IoT is to develop things in a lightweight and smarter way. The designing of sensors in IoT is different as compared to commonly used sensors. There are so many sensors available according to their need and design. In this chapter, the authors have explained the methods of IoT, wireless sensors, application of IoT in healthcare, its challenges, its requirements, and the various sensors used in this method.

With the IoT, nowadays it is considered as one of the most important, newest, and fastest spreading mechanisms in the communication area. IoT consists of devices and integrating sensors at daily smart objects that are linked to the internet through wireless sensor networks (WSNs) that lead to opening the door to new methods of exchanging the data, which were not possible before. IoT has several applications, and one of the most effective areas is E-health or smart healthcare. In this application, radio frequencies based on wireless networking technology and wearable sensors are connected to the base station [2]. Currently, a great number of researchers are learning IoT applications in the E-health field. E-health terms had been recently established, which handles management of healthcare with the backing of electronic communication and processes techniques. E-health systems, such as wearable devices and cell phones, provide continuous monitoring of patients. Thus, this will provide many advantages like cost saving, transportation, insurance costs, and services of healthcare providers. Therefore, this will lead to achieving the goal of facilitating secure interactions among healthcare providers and patients, which will lead to better quality of healthcare, and save the time of patients [3]. E-health applications are a point to hack data, and increasing issues of security aspects are rising in numbers, and because critical data are sent through E-medical records, also because of the growing of user-wearable technology [4]. So, the main concern of IoT is the high level of security that is needed to keep all the communications secured. The concerns of security are extended due to the rapid deployment of IoT [5]. At IoT, security is mainly part of E-health applications, which provide a high level of security for medical data [6]. There is a need for more efforts and more researchers to handle the security problems. However, many of the researchers had searched about an open issue at IoT [7]. Researchers aim to meet the requirements for making security a major factor to build IoT E-health applications to protect data communication mechanisms [8].

2.1.1 Internet of Things

The IoT is the most effective area of research, where sensor nodes and smart devices can collect the information from different sources and communicate with the server without human involvement. In IoT, the main important concept is WSNs, in which data are shared and communicated with the help of sensor nodes. These sensor nodes are distributed randomly in the specified areas for collecting and sensing the information of different parameters. The IoT is a combination of nodes and sensors known as "things." It refers to a node, device, or sensor that measures the physical quantity and converts it into the digital quantity [9]. It includes many applications like agriculture, military, home office automation [10], E-health, weather forecasting, monitoring and controlling systems, etc. Global Positioning System (GPS) works on the same technique and can be the solution to all of these applications, but the cost of GPS is very high and it consumes very high energy power. So this area has many numbers of researchers who are looking to optimize these services. This is directly involved in collecting and monitoring the information of sensors over the cloud internet. The important components of IoT are radio-frequency identification (RFID) and WSN. RFID gives the permission of automatic identification and capturing of data by using radio frequencies through tags and a reader. These tags are very powerful as compared to the traditional barcodes [10]. IoT is useful and involved almost everywhere. There are a few areas of applications shown in Figure 2.1(a).

2.1.2 Wireless Sensor Network

WSNs are the collection of distributed sensor nodes that can be used for multiple applications, like E-health, smart agriculture, military, weather management systems, environmental controlling systems, etc. [11]. Sensors are the main components of WSNs and they are distributed in an ad hoc manner. Sensors include sensor subsystem, processing system, and communication system. Every sensor has a base system and connects with the internet for communicating and sharing the information. There are many routing algorithms designed according to the requirements of the system. Each routing protocol is different from one another and has its own applications and limitations; it can be used according to specific work [12]. In Figure 2.1(b) shows the structure of WSN.

2.1.3 Body Area Sensor Network

A WSN contains an associated field known as a wireless body area sensor. Once the WSN has reached at a high level of development, the latest incoming wireless body area sensor network has inherited its place. These techniques become possible due to its small size components and technological enhancements. This technology gains the recognition due to the up gradation of the technical event and easy-to-use wireless technology. The healthcare sector has provided this technology with significant attention in recent days [13]. The microsensors are the driving force of this technology. The microsensors could easily be placed on the human or physical body to monitor the various physiological parameters. These parameters could be sent to other devices for the further analysis purpose or for taking any decision on the output. This technology is a valuable asset for the healthcare sector because it could be used for the disease diagnosis and could also function as an alarm system for such complicated diseases for the precaution purposes. The intent of this study is to present the various state-of-art aspects of the wireless body area sensor and its architecture, its framework of programming, security concerns, its routing protocols, and its application. Figure 2.1(c) depicts the framework of body area sensor network.

2.1.4 Smart Healthcare

Though the modern healthcare system is fully equipped, still it is facing the overburden due to the steady increase in the elderly population and development of chronic

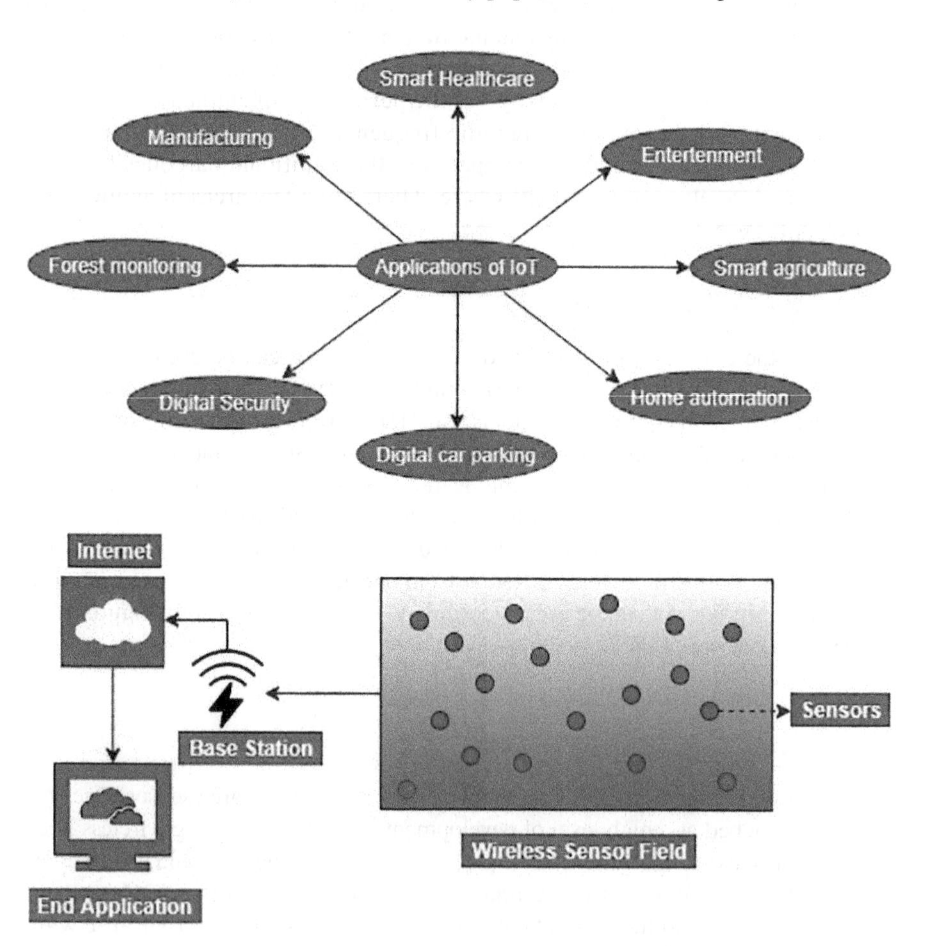

FIGURE 2.1 (Continued)

disease conditions from the acute stage conditions. This situation has caused the high demand of the medical professionals throughout the globe [14]. The healthcare practitioners are looking for a solution to overcome this burden to facilitate the quality services to critical patients through the remote location using the technology.

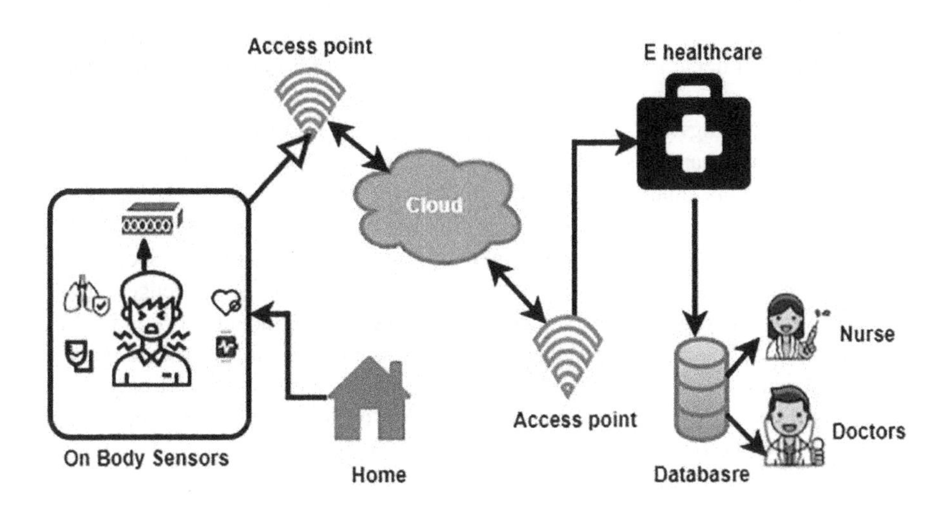

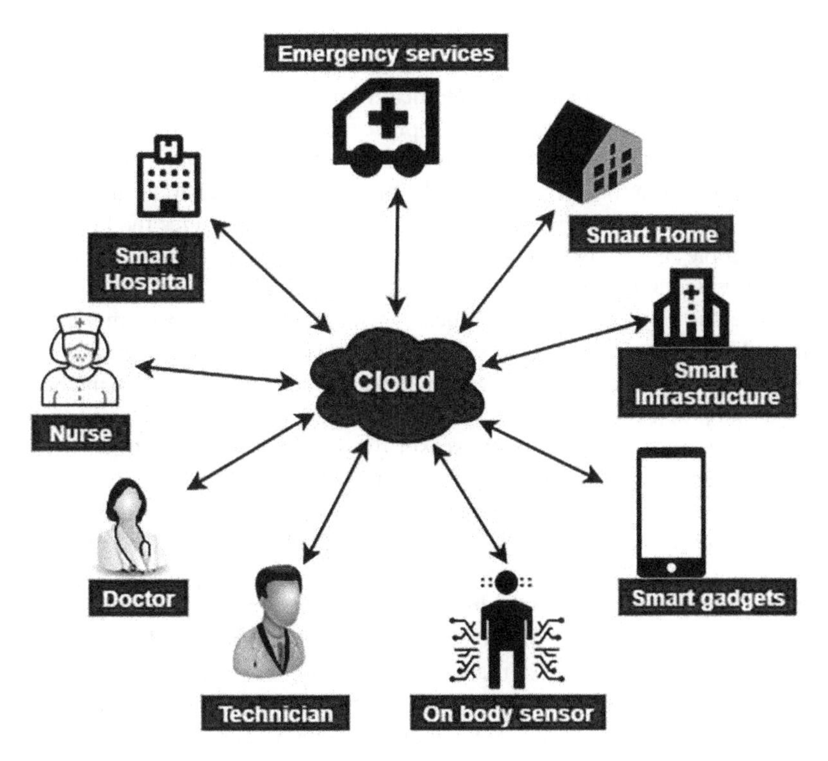

FIGURE 2.1 IoT, WSN, BASN, and smart healthcare.

The new technology of this era known as IoT came in knowledge to cope with the solution from this problem. IoT became the area of research for the various aspects in the recent days. The thrust area of this study is mainly focused on the patients suffering from the diseases like diabetes, Alzheimer's disease, and Parkinson's disease. Another part of research is mainly focused on the rehabilitation process by which the monitoring of the patients can be performed for the specific purposes. Figure 2.1(d) [14] depicts the process of internet connections supporting the healthcare mechanism.

This chapter brings out the exhaustive survey of IoT and IoT in healthcare [15]. The scope of this chapter is to provide a brief idea about IoT, its application, architecture, components, challenges, and the IoT-based healthcare system.

The remaining chapter structure is as follows: Section 2.2 includes a detailed related work of IoT, body area sensor network, and smart healthcare. Section 2.3 includes a detailed survey of IoT, like its architecture, components, challenges, applications, sensors, and many other related things. Section 2.4 includes IoT in healthcare; in this section the authors discuss how IoT is involved in healthcare, how it reduces the complexities of traditional healthcare, which technologies are important in IoT-based healthcare, and what the limitations of smart healthcare are. Section 2.5 discusses about the future scope, and Section 2.6 concludes the work.

2.2 RELATED WORK

The advancement of IoT provides the reasonable methods for the remote health monitoring. The mechanism of remote health monitoring provides the method of monitoring non-critical patients at their place, which helps in the reduction of the pressure from the hospitals and the medical practitioners. This mechanism plays an important part in the life of persons belonging to the rural areas and those who lack access to the healthcare facilities. This technology provides the better option for such people or elderly people to provide better healthcare facility at their end. IoT has provided a new opportunity to the persons to take care of their own health and reduce the burden on the healthcare resources in the short time span [16]. Every technology comes with some drawbacks, and the main drawback of the IoT technology is the safety risk. The remote health monitoring system is incapable in the situation when the sensor loses connection or the patient moves out from the cellular range or the battery of the sensor drains out. The other drawback is the database security and data sensitivity, because all the data get stored in the single database [17]. All the drawbacks of the IoT could be addressed and the proper solution could be found out. This study has highlighted such methods throughout the work.

In [18] the authors state that the IoT may be a platform where devices, processing, and communication became smarter, intelligent, and informative. In this paper, the authors give a concise discussion about the architecture of IoT and the connected scenarios for different physical devices for creating a network. This paper also gives the idea about the related tools, technologies, methodologies, and the requirements of the developers.

In [19] the authors indicate a network in which physical objects or devices are able to collect and share electronic information. The IoT network contains a wide set of smart devices, sensors, and actuators, which transfer data over the network through IoT applications. The extension of IoT applications and wireless technologies is leading to a new opportunity for the growth of various important fields, known as enterprise, education, agriculture, transportation, and a very important field, namely healthcare. In this paper the author discusses the importance of IoT, as well as the security issues and authentication of connected objects and exchanging data. The author proposed a secure and authentication scheme with group nodes for IoT-based smart healthcare application. This paper also focuses toward the proper energy utilization of devices or sensor nodes.

In [20] the authors have explained in the research work about the outfitted sensors, which are connected to each other for better communication. This method provides the result in a simultaneous manner. The authors also highlighted the thought about the technology to make this technology more robust and powerful. They have also discussed the exhaustiveness of the technology and its broad range sensors and its applications.

In [21] the authors have talked about the quick enlargement of IoT technology that provides the connections among the different smart objects through the internet, and these connections provide interconnection among the objects to gather more data to provide better healthcare facility. The authors have also discussed the semantic data model and proposed the method that discussed about the storage and interpretation of the data obtained by the interconnections of the model.

In [22] the authors have explained the method using the concept of cloud computing, that provides the method of handling the massive data obtained through the various interconnected sensors. This helps in the sharing of information through the IoT network to make healthcare service more reliable. The authors have also explained the newly developed method of cloud computing, along with the IoT technology to provide better healthcare services.

In [23] the author explained about the wireless body area sensor network. This is a subfield of a WSN. The author had also discussed about the architecture of communication networks, its applications, and the security issues, along with the routing protocols with energy-efficient functions. The microscale of sensors provides the better output for the modern healthcare.

In [24] the authors explained the energy constraint of the IoT network. They have sketched out the narrowband IoT (NBIoT) for the healthcare sector. They have also provided the various implementable proposals for the NBIoT for the healthcare sector. The designed protocol was made compatible with the Long Term Evolution (LTE) platform. The NBIoT faces the main challenges related to the security concern, but the authors have handled this problem very smoothly and came out with the solution of this problem.

In [25] the authors have shed the light on the healthcare monitoring system on the issue of elderly populations in the community. The authors have explained point-wise issues of the IoT system, including the standard architecture model and its application, along with the future and scope of the IoT in the field of healthcare sector.

In [26] the authors have defined the BSN (body sensor network) technologies for the designing of the small-scale and light-weighted sensors for the better diagnosis of the patients located at the remote location. This could be possible by the proposed method, which helps in the monitoring the various parameters of patients connected with the sensors. The authors have discussed in their paper regarding the BSN-based modern healthcare system. The method of BSN care had been proposed, which provides the secure IoT-based healthcare system for the better outcome in the field of healthcare monitoring system.

In [27] the authors have discussed about the mobile medical devices, which are feasible to take decision according to the need. The device is equipped with the biomedical sensors, cloud computing methods, and the concept of big data analysis to manage the E-health monitoring system. The authors explained the new methods of security scheme for the E-healthcare management. The schema provides the mechanism in which a local station is set up, which connects with the hospital server, and both the networks authenticate the obtained data and secure the collection of healthcare data. The main feature of this proposed schema is lightweight, and secure from the different types of attacks.

In [28] the authors demonstrate the communication between the smart devices and the mobile, which is android-based. The devices are connected via the Bluetooth network and the data stored on the cloud. This paper discusses about the method that helps in the catering of the admitted patient's data on the mobile devices. This proposed scheme helps medical practitioners to communicate with several patients at the same time, which reduces the physical burden from the healthcare system.

In [29] the authors have proposed a method in which the physiological parameters could be monitored after every 10 seconds. The connected sensors collect the various physiological parameters, such as body temperature, pulse rate, heart rate, blood pressure, and ECG signals, etc. Once the data have been collected, the sensors analyze the data and resolve the errors if any, and they also replace the missing values if present. The signal is sent to the control system and also sends a message to the doctor immediately once after the sensors detect any problem. The major advantage of this system is the consumption of less power supply, high-speed process of communication coverage, and the security and privacy of the patients' obtained data. The secure IoT communication is highly encrypted with an advanced encrypted technique.

In [30] the authors discuss about the superfluity of wireless body area network (WBAN) applications and network construction. Data collection, data transmission, and data analysis are the important phases of IoT. This paper also classifies the purpose of routing protocols and its use according to the use.

In [31] the privacy and security of the patient's data is a main concern in the field of IoT. The author had explained the design, which is highly secured and guaranteed to the seclusion of the patients' information, and also proposed the method of exchange of data from the clouds. These secure methods help the healthcare centers and the doctors to manage the data securely from the remote location and provide the better healthcare. The data can be saved and retrieved from the regular basis from the cloud and the healthcare centers. This approach provides the doctors a new method to check the regular health status of the patients and can provide the better solution for the problem and can also check the progress of the patients.

All these research studies have shown that IoT is very much successful in the healthcare sector and provides an enhanced version of E-healthcare. Still, there are many limitations.

2.3 OVERVIEW OF INTERNET OF THINGS

The IoT technology provides the process of the complete automation for the analysis and integration of the obtained data. The IoT used the technology of sensing, the better the sensor the better the result, because the complete processing data are dependent on the sensing system. The sensing and networking and robotics made this technology more robust [32]. IoT technology itself includes the design of the device architecture, its layers, and many more. Before applying the application of IoT, the users should be aware with the infrastructure of IoT as shown in Figure 2.2.

2.3.1 Architecture of IoT

The architecture of IoT is a five-layer structure, where each and every layer has its own working and prototypes.

- The perception layer is as same as the physical layer of the Open Systems Interconnection (OSI) model. This layer works on gathering the data and physical objects.
- The network layer deals with the network technologies and transfers the collected data throughout the system with the assistance of different sensing or physical devices.
- The middle layer of IoT is also called the perception layer. This layer helps in the storing, analyzing, and processing of the huge amount of data

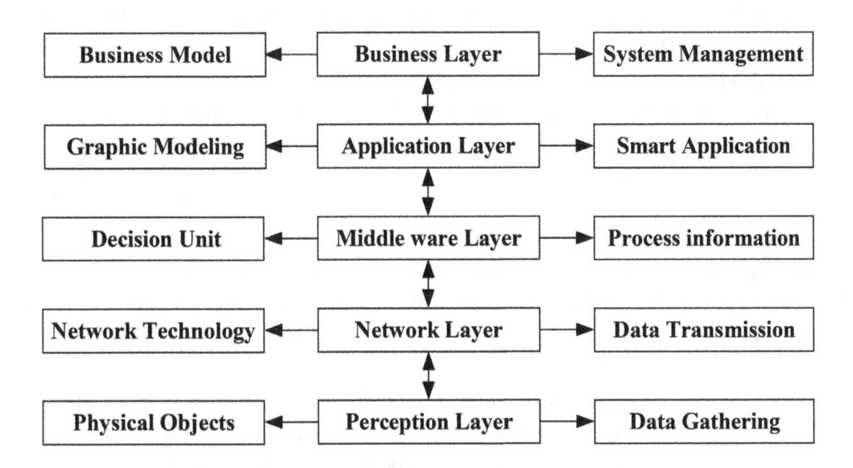

FIGURE 2.2 Layering architecture of IoT.

obtained from the application layer. This middle layer is capable in managing and providing a vast set of services for the lower layers. This method includes various technologies such as cloud computing, big data analysis, and database management system.

- The application layers focus toward smart application or e-application, which manages the bridge between middleware and business layer.
- The business layer completes the setup of the IoT system, including its application, its business mechanism, and profit model and user's privacy.

2.3.2 IoT Key Features

There are many significant features of IoT, which includes artificial intelligence (AI), connectivity between different devices, sensors, actuators, and some other small devices.

AI: The IoT converts everything into "smart" things. It means enhancing the authority of data collection, data analysis, data monitoring, AI algorithms, networks, and infrastructure. AI is the backbone of IoT as it can design a platform for proper working.

Connectivity: Another important feature of IoT is connectivity because a number of devices can perform if and only if they can connect properly. Connectivity may be short-range connectivity or long-range connectivity. There are many devices, which establish the connection between different devices.

Sensors: These are the tiny devices in IoT networks and the overall networking of these networks passes through these sensors. There are different sensors for every different task. The main function of sensors is to sense the data and collect it for different sources and convert it to the physical quantity.

Actuators: These are the displaying devices. These components are machines, which can monitor, control, and display the measured data or quantity. They actually convert energy into the motion and perform specific functions. There are different types of actuators.

Small devices: These devices are predicted, tiny, cost-effective, powerful, and more sensible over the time. The overall agenda of IoT is to design the small and lightweight devices as compared to the traditional networks.

2.3.3 IoT Advantages

The advanced use of IoT directly indicates the advantages over traditional networks. The advantages of IoT are in every area of human life. A few advantages are discussed below:

Improved end-to-end interaction: In the IoT, the interaction between client and server increases dynamically. Current analytics focus toward accuracy, fast, correctness, and engagements. IoT completely transforms this to achieve richer and more effective engagement with users.

Optimization of technology: This is a similar technology, which improves the experience of customers, which improves the use of the device and helps the technology to improve the application. IoT provides the opportunity to solve the critical field data.

Include more devices: IoT is not limited to specific ones; it opens the arms to a larger number of nodes and includes more devices to make them function smoothly.

Enhanced data collection: Traditional data collection techniques are very complicated and time-taking. IoT makes it simpler and improved. Every sensor directly sends the sensed data to the access node, and then the access node sends it to the base station. Duplicated or multiple nodes are not present in between.

Improved data monitors: Data monitoring in IoT plays an important role; no human involvement makes it more powerful and precise.

2.3.4 IoT Disadvantages

IoT became a very important part of coming years. It reduces the human efforts at a very high level, but still there are many limitations present, and many researchers are working in these areas and trying to make this network more efficient. A few limitations are given below.

Security: In IoT, security is a very critical issue. To create a communication, many devices are involved in this and generate a huge amount of data over cloud. A small gap of security control creates a big security threat and hackers can misuse the system.

Privacy: IoT systems are still not very secure in terms of private data. Single information travels over the multiple channels and creates many replicas as it may interrupt the personal information of any user.

Complexity: IoT systems are complicated in terms of system design, device deployment, data maintenance, data controlling, and monitoring. More devices sometimes create more complex situations.

Cost: Tiny sensors are very costly and they may increase the overall cost of the network. IoT devices are different from the regular devices and the cost of these devices is higher than old devices.

Data storage: In IoT infrastructure, everything is connected to the cloud and it produces a very large amount of data. So, it will deal with the efficient use of that storage and manage Big data.

2.3.5 Components and Connectivity of IoT

IoT architecture uses components for interaction and communication with other IoT devices. They need components for processing, internet interaction, handling the web services, accessing the other services, and many more. At every phase it needs different components and connectivity [33]. There is connectivity according to the distance and use. Table 2.1 shows the components and connectivity types.

TABLE 2.1
Connectivity Type and Its Function

Connectivity Type	Function
IoT LAN	Short-range communication and connecting local buildings and organization to the internet.
IoT WAN	Connect a variety of networks in large geographical area.
IoT Node	Connect to another node inside a LAN or connect to the internet to a WAN.
IoT Gateway	A router connects LAN and WAN to the internet. It can connect several LAN and WAN for data transferring.
IoT Proxy	Perform application layer functions between different IoT devices.

2.3.6 Sensors and Actuation

A sensor can sense the data from a source and change it into the physical quantity. It can collect the data and send it to the above layer for further processing [34]. Sensors perform as input devices and perform some input functions by sensing the physical changes. For example, electrical signals generate heat with the help of a temperature sensor. Atmospheric pressure is converted to the electrical signals by the use of a barometer. It is sensitive toward the measurable property (e.g., a temperature sensor senses the ambient temperature of a room) and insensitive to other measurable properties like heat or light (e.g., a temperature sensor does not bother about light or pressure while sensing the temperature). It only can measure the data or sense the data, but it cannot directly perform the others' functions (e.g., measuring the temperature does not reduce or increase the temperature). Table 2.2 describes the sensor types and its function.

TABLE 2.2
Sensor Types and Function

Sensor Type	Functional Devices
Light sensor	Light-dependent resistor and photo diode
Temperature sensor	Thermocouple and thermistor
Force sensor	Strain gauge and pressure switch
Position sensor	Potentiometer, encoders, and opto couple
Speed sensor	Reflective opto couple and Doppler effect sensor
Sound sensor	Carbon chemical sensor and piezoelectric crystal
Chemical sensor	Liquid chemical sensor and gaseous chemical sensor
Proximity sensor	Inductive, capacitive, photoelectric, and ultrasonic sensors
Smoke, Gas and Alcohol sensor	IR sensors and Infrared LED
Infrared sensor	Radiation thermometers

Actuators are a device that can convert the electrical energy into some useful energy. The actuator is the parallel components of the sensors. Sensors are responsible for sensing the data, and actuators are responsible for displaying the sensed data in the exact form. It may be a machine, system, or device that can move or controls the processing or the system. An actuator works with the control system and requires control signals and environment energy. After receiving a control signal, the actuator reacts to convert the energy into mechanical actions [35]. The control system for actuators may be a preset mechanical, electronic, or thermal system, or it can be a software-based system, (e.g., a printer, driver, and robot) or a creature, or any other input device.

Types of actuators:

- Hydraulic
- Pneumatic
- Electrical
- Thermal/magnetic
- Mechanical

2.3.7 Applications of IoT

In today's world, the IoT is everywhere. Almost in each and every field, the IoT plays an important role and makes the previous system smarter and more effective. Few are shown as Business/Manufacturing (it is a real-world analytics of supply chain, equipment, robotic, and machinery), Smart Healthcare (flexible healthcare monitoring, controlling by electronic record keeping, and managing the pharmaceutical for safety and precautions). Devices connect to hospitals, doctors, and relatives to alert them of medical emergencies and take preventive measures, retail (digital transitions, Smartphone shopping, analysis of the choice of customers), security (face and thumb recognition, touch locks, voice matching sensors, etc.). Digital locks are also the part of security; it is mostly useful in Smartphones and digital door locks and it can be used for locking and opening the doors remotely. The system head or owner scans alteration key codes and locks quickly to maintain or control the access of users. There are many more areas, which are using the application of IoT for making the function more smart, digital, and appropriate. Figure 2.3 shows the share of IoT in different sectors.

2.3.8 Comparison Between Traditional and IoT Network

The very initial difference between IoT and old networks are the interaction of humans. In IoT, human interaction is very less because every work is designed by the machines. Here users don't need to request the system; the system automatically triggers the monitor from time to time. It also generates a big amount of useless data, so to manage big data is a very required task. The IoT system is not capable of working with traditional heavy algorithms, and it can work on lightweight environments. As compared to the old system, IoT systems are more active, so that it consumes more energy and needs more security. Table 2.3 shows the comparison of traditional and latest IoT networks.

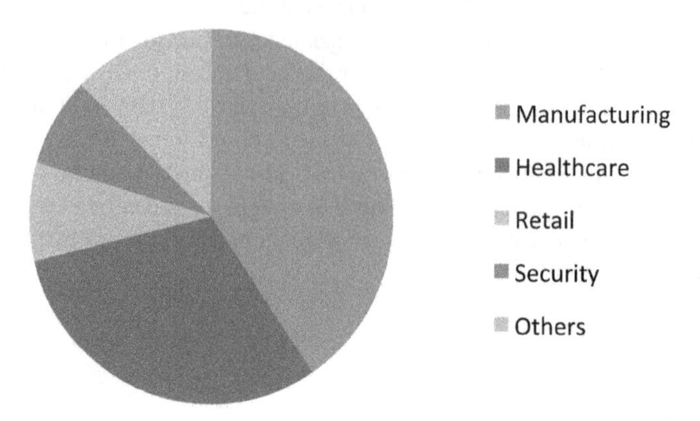

FIGURE 2.3 Market share of IoT.

TABLE 2.3
Comparison between Old and IoT Systems

Parameters	IoT Infrastructure	Old Infrastructure
Creates by	Machines	Human
Security	Lightweight security	Heavy securities
Cost	Low	High
Sensors	Small in size but more in number	Large in size but less in number
Interface	More	Less
Database	Lightweight	Heavy
Operating system	Lightweight	Heavy
Energy	Consume more energy	Consume less energy
Sleep frequency	Mostly active	Sleep when no use

2.4 INTERNET OF THINGS IN HEALTHCARE

E-health term had been recently established, which handles the management of healthcare with the backing of electronic communication and processes techniques. E-health systems, such as wearable devices and cell phones, provide continuous monitoring of patients. Thus, this will provide many advantages like cost saving, transportation, insurance costs, and healthcare providers. E-health applications are an exhibition to hack data and increasing issues at security because of the maximum number of multilayer points and multi-gateway processing. The essential and critical data of the client move through the e-channel and produce a replica on cloud. Replication can violate the security of the system [36]. This phase includes the model of the IoT healthcare, wearable devices, communication channels, cloud computing, and security aspects.

2.4.1 Model for IoT Healthcare Systems

The IoT-based healthcare system had a broad range of applications, which inspires the researchers to design such a model, which is apparently required for the modern-day world. All the previous research focuses on the monitoring system for the healthcare in regard to the wireless and extremely wearable sensors, and these sensors are functioning according to their system requirements [37]. The task in the IoT-based monitoring system is to keep the size of sensors at microscale level to provide the complete comfort to the patients and facilitate the process of the patient health monitoring system. If the plant-able sensors and cameras could be implanted in the patients, the patients will become more receptive than they were before. The flow diagram of IoT-based healthcare monitoring system is depicted in Figure 2.4.

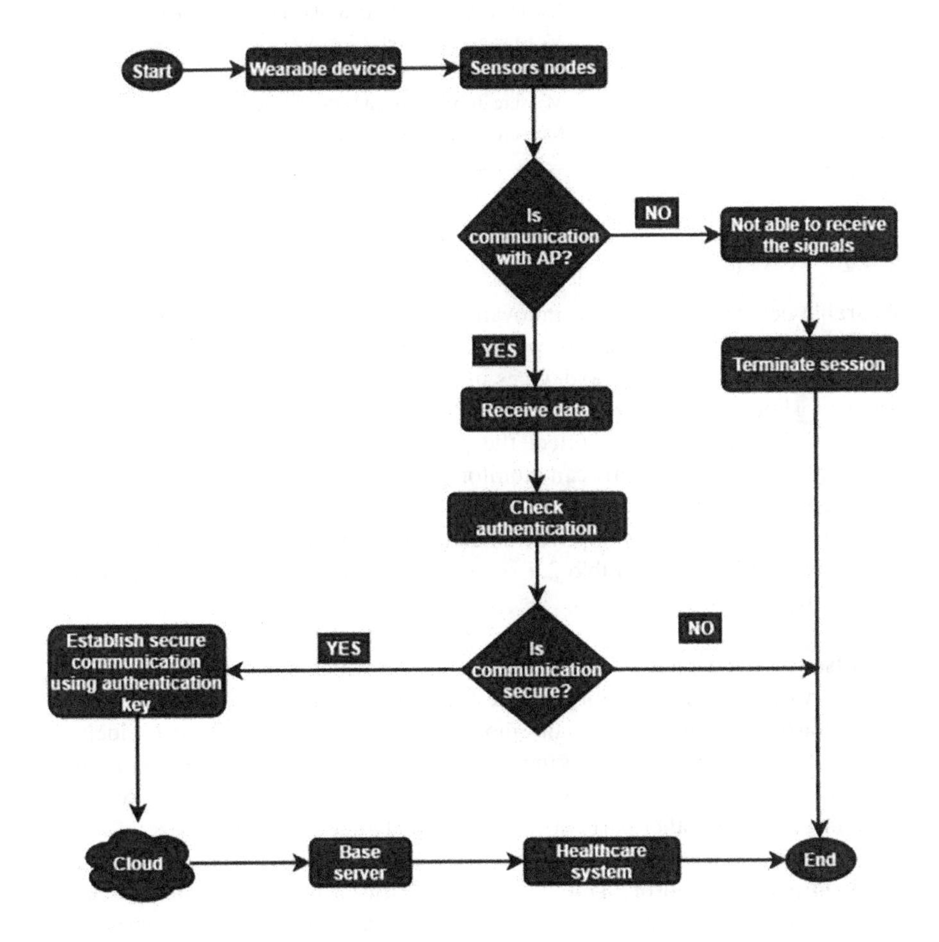

FIGURE 2.4 A model for IoT healthcare.

TABLE 2.4
List of Medical Sensors and Their Usage

Name of Sensors	Uses
Heart rate sensor	To sense the heart rate and send it to the monitors.
ECG (Electrocardiography)	To monitor the electrical function of heart and this transfers the status of heart and the function of its muscular contractions.
EMG (Electromyography)	To measure muscular activity and transfer the electrical signals for gesture recognition.
EEG (Electroencephalography)	To capture the signals for brain activity.
Blood pressure (BP)	To observe the pressure of blood in the heart system.
Respiration rate	To count the breathing rate.
SpO2	Calculate the amount of oxygen which dissolves in blood.
Skin conductivity	To measure the moisture level of skin.
GSR (Galvanic skin response)	Perspiration of skin type.
Co_2 gas	Calculate the level of carbon dioxide from mixed gas.
Glucometer	Measure glucose level.
Motion sensors	To trace steps count, sleeping and burn calories.
Stress sensor	Measure the pressure on body and brain.
Accelerometer	Measuring the energy consumption.

2.4.2 Wearable Devices and Medical Sensor

Wearable devices are the most innovative idea for IoT networks. These devices can interact with the patient as well as the doctor without any physical presence. These are the electronic devices in healthcare that can be worn, like smart watches, shoes, etc. These devices are designed to collect the data of patient health [38]. These devices collect the patient data and also transfer it to the server so that doctors or staff can monitor this data. If any abnormality is present in the collected data, the doctor can initiate a checkup or make an alert call to the patient. There are so many wearable and medical devices available for making E-health easy to use. Table 2.4 shows the list of wearable devices and their functioning.

Pulse sensor: This is a most important sensor, which helps in the detection of various critical conditions like heart attack, vasovagal, syncope, pulmonary embolism, and many more conditions. This is the most widely studied sensor for the tracking of the medical history and tracking of the fitness of the patients. These sensors can be implanted on the chest, wrist, fingertip, and other parts of the patient body due to the small size of the sensor.

Blood pressure sensor: This is also a vital sensor, but the use of this sensor is limited because it measures only the blood pressure. Both types of blood pressure could be measured, i.e., high blood pressure and low blood pressure. The high blood pressure is the main risk factor for the cardiovascular diseases, along with the risk of heart attack.

Heart rate monitoring: The heartbeat of the patients could be obtained by placing the patient finger on the heart rate monitoring sensor. The sensor predicts the blood flow through the finger placed on the sensor. Sensors are set up with the standard light resolution system, which contains both high- and low-intensity beam for the calculation of heart beat count. The sensors are connected with the Raspberry Pi, which are capable to calculate the beats per minute (BPM).

Respiratory rate sensor: This sensor is capable of monitoring the various functions of lungs, such as respiratory rate, or the breath inhale or exhale per minute. These measured observations help the sensor to monitor any critical condition which can occur in the future. These conditions are asthma, apnea episode, lung cancer, tuberculosis, obstruction in the air passage, and panic attacks, or many more.

Body temperature sensor: This sensor helps in the monitoring the person's body temperature, which could be used to detect the condition of pyrexia, heat stroke, and other symptoms related to body temperature. The body temperature plays an important role in the diagnosis of any infectious disease because fever is the response of any uncomfortable conditions of a person's body.

Oxygen meter: It is named as pulse oximetry sensor, which predicts or calculates the amount of oxygen in the person's blood. The blood oxygen level may not play a pivot role, but it works as an indicator for the respiratory functions and can also help in the diagnosis of the conditions of hypoxia or hyperoxia. This makes the oxygen meter a valuable asset in the field of healthcare sensing.

Sugar level: In the recent world the number of diabetes patients is increasing drastically, which enables the glucose sensor as a valuable asset in the medical healthcare service. The glucose sensor monitors the blood glucose level and the required sugar for the biological metabolic and catabolic mechanism. The sensor also predicts the glucose concentration, which is generated, and breaks down through the biological process. The sensor is selective and sensitive toward the function, and in the recent days sensors are based on the enzymatic glucose reactions occurring in the biological system.

Motion sensor: The motion sensor predicts the movement of the object. This sensor is capable of the measuring or calculating the number of steps walked by the person and enables the fitness of the person, or this may also help in the maintaining the fitness of the human being.

Other wearable sensors: Along with the above discussed sensors, there are several sensors available to help in the healthcare sector to promote the quality of the human health, and these sensors help in reducing the burden from the medical practitioners. The sensors have the capability to sense the data and send to the control room or doctors to take up the decision from the remote location.

2.4.3 Communication Between Devices

In the IoT environment, many devices are connected with each other, and every device is well designed for its works. These devices need to work together and communicate with other devices for collecting or transferring the sensed data. Few devices are placed very close because they need short-range communication, but some devices are placed very far, and in that case short-range communication is not suitable. It may drop the data, so here long-range communication protocols or tools are useful. Short-range and long-range communications are basically decided by the distance. IoT devices have limited battery, so to make proper communication is a very challenging point, according to the need to choose an appropriate tool for communication. The type of short-range and long-range communication is discussed below.

2.4.3.1 Short-Range Communication

With regard to wearable human services frameworks, short-extend interchanges are frequently utilized between hubs, especially between sensor hubs and the focal hub where information preparation happens. Short-go interchanges norms can be utilized for different purposes (e.g., creating network systems for keen lighting). This study centers around the reason for building up a little WBAN that is involved with just a couple of sensors and a solitary focal hub [39, 40]. Some short-go interchanges norms exist, however; maybe the most normally utilized terms in IoT are ZigBee and Bluetooth Low Energy (BLE). This area advance examines these norms and suitability for the usage of IoT human services framework.

Bluetooth: It was created by the Bluetooth Special Interest Group (SIG) to give a vitality of effective uses and requirements that can be utilized by battery (coin-cell) worked with different gadgets, wearables, and objects. It was additionally intended to empower IoT, interfacing little fringe gadgets to preparing gadgets, for example, advanced mobile phones. Bluetooth technologies are designed by star topology, which is mostly reasonable for medicinal services applications or in healthcare sectors [41]. The central hub works as a central point for the created star topology, with sensors and other objects. The sensors will have no compelling reason to speak with one another legitimately. The defined area of working for BLE is 150 m in an open area, and it may be reduced in a nonperfect environment. It additionally has a minimum range of 3 ms and a high information pace of 1 Mbps. The range of BLE is unmistakably adequate for some use in human services. WBAN is an effective area of medical science, which is basically working with the use of Bluetooth technologies [42]. This short-range application is very much effective and useful when sensors and devices are placed close to each other, and they can easily communicate with others. IoT healthcare mostly uses this short range of communication to create local infrastructure.

ZigBee: Another short-range technology is ZigBee, which was planned by the ZigBee Alliance, explicitly for giving ease to use and less consumption of energy in M2M interchanges. This expands further in the standards of IEEE 802.15.4. It is ordinarily recognized for machine's work systems. This technology also uses star topology expected of a wireless body sensor area network. This technology selects one of the devices as a focal hub and others as detecting hubs. Distinctive ZigBee modules give various qualities as far as range, information rate, and force utilization [43]. The easiest ZigBee has a scope of up to 30 m in an urban domain, and yields just 1 mW of intensity for transferring and receiving. The latest version of ZigBee is ZigBee Pro. It has a larger scope of 90 m transmission in similar conditions and it can transmit with 63 mW power. Then further the ZigBee Pro reaches up to 900 XSC and 610 m in an urban domain, with 250 mW of intensity being utilized to transmit. There are ZigBee-based answers for a wide assortment of uses; however, for the utilization instance of a social insurance WBAN, the XBee 1 mW would be reasonable. Just a little range is required for on-body correspondences, so picking the most reduced force arrangement is best. As we already discussed, these sensor nodes are battery-operated, so power consumption is always an essential; issue in this sector [44]. To choose the right tool for communication increases the life of digital healthcare, which automatically improves the performance of the overall system. In IoT many local areas are connected with a wide area network, so we need almost every type of communication tool according to the use. A big challenge in IoT healthcare is also to choose the correct technology at the correct place.

2.4.3.2 Long-Range Communication

Low-power wide-area networks (LPWANs) are a subset of long-run interchanges measured with high appropriateness for IoT applications, mostly in healthcare. The scope of an LPWAN is commonly a few kilometers in an urban situation. That is altogether longer than the scope of conventional IoT interchanges types, for example, Bluetooth or ZigBee. These reaches are in the request for meters, and along these lines, would require broad and expensive work systems administration or would be conceivable for human services. LPWANs additionally have a huge favorable position over cell systems, for example, 3G in that they are intended to help short explosions of information rarely [45]. The IoT healthcare basically depends upon the door-to-door communication, so to maintain this quality is a big issue for this system. This is reasonable for an enormous number of human services applications, includes checking general well-being and accepting hourly updates, observing basic well-being and getting crisis calls, monitoring the reports, and controlling the abnormal identities and restoration where updates may just be important once day by day. This structure standard additionally takes into account a low-power gadget plan, which thusly guarantees that the planned social insurance gadgets will work for longer before human connection is required to energize or change battery systems [46]. This decreases the danger of patients while disconnected and gives more comfort to the customers. In this point of view, these crucial points are recommended that LPWANs are the best solution for communication between patient and doctors for further processing over the cloud-based infrastructure. There are many technologies in long-range communication that are discussed below.

SigFox: Perhaps the least difficult of the LPWAN norms, SigFox gives restricted usefulness; however, it is generally sent contrasted with different measures recorded. This technology is created with the help of OSI model initials layers. It has base stations and sends and receives the signal over the base stations. SigFox also utilizes well-known Star topology [47]. The hubs and the central devices are intended to be uplinked to improve battery proficiency. It is feasible for a hub to get a downlink, yet it should expressly demand it. As an affirmation of receipt is significant for well-being information, the downlink would need to be mentioned. Tragically, a restriction of SigFox is that the downlink must be mentioned four times each day. SigFox has a high system limit and can bolster roughly 50,000 hubs with a solitary door. This is tantamount to NB-IoT's 52,547 hubs, which has been demonstrated to have the option to help 40 gadgets for each family unit, expecting a family thickness proportional to London's and a buried site separation of 1732 m. One very much situated base station would empower each occupant of the area to be associated with social insurance suppliers by means of SigFox. This is noteworthy for human services applications, particularly in provincial regions. This provides the facility for uplink and downlink, so we can manage the energy as per the use. As in healthcare, energy is a crucial factor for maintaining the flawless communication, and this technology provides a better result in healthcare and also manages a long-range communication without any break.

LoRa and LoRaWAN: Technical data of the most popular terms LoRa and LoRaWAN are introduced and composed by the LoRa Alliance. These are the protocol of wireless IoT in a wide area network. They are useful for IoT connectivity and communication when objects are in long distance. These technologies play a very useful role with IoT gateways, IoT platforms, cloud computing, and business layer performance. Generally, these networks are mixed technologies networks; for example, in IoT healthcare few devices are closely placed and few objects are long distance, so closely placed objects use different technologies and long placed objects use different technologies. LoRa and LoRaWAN are the wide range technologies as they are majorly working in noncellular form and are widely used in smart healthcare, agriculture, environment monitoring, and many more. This section discusses the key segments of the standard dependent on this source, and along these lines, the intrigued peruser alludes to it for additional data about LoRa and LoRaWAN. LoRa is a physical layer convention that uses twitter spread range strategies over a wide data transfer capacity of at any rate 125 kHz. This gives low-power, long-extend correspondences with high flexibility to deliberate or ecological impedance. LoRaWAN is based on the LoRa standard, in the system layer. It uses a star topology, and hubs are non-concurrent; they possibly convey when they have to, for example, after an occasion or planned estimation [48]. Planned messages from hubs would suit long-haul observing applications, while occasion-driven messages from hubs would suit crisis checking. This technology is very much successful in IoT healthcare because of its different capacities and different bandwidth.

Narrowband IoT (NB-IoT): In the coming 3GPP Release of 13, NB-IoT works in the authorized groups of GSM or LTE and gives longer-extend less-power correspondences. As NB-IoT has been created dependent on LTE, a great part of the current LTE equipment can be utilized to convey it quickly and adequately. There are three distinct manners by which NB-IoT can be sent, permitting simpler conjunction with existing systems. These arrangement modes are in-band, monitor band, and independent. NB-IoT provides better services in the coordination between sensors, and it is economical as compared to others. It is simpler in use and performs the functions effectively. Due to larger demands of healthcare, NB-IoT utilizing in-band mode includes holding LTE Physical Resource Blocks (PRBs) from the current LTE organize. In monitor band mode, NB-IoT uses the data transmission of a current LTE bearer's watchman band. At last, in solitary mode, GSM transporters can be re-cultivated and utilized for NB-IoT, or NB-IoT can exist in a totally new data transmission [49]. NB-IoT is a longer-band technology, which provides economical and much effective in emergency situations. As per the demand of healthcare, cost, time, speed, and efficiency are the important issues, and NB-IoT is much successful for fulfilling the requirements of IoT healthcare. It can provide a better communication and handle the task effectively.

2.4.4 Radio-Frequency Identification

RFID, the versatile innovation, encourages and helps with improving the uses of IoT human services. RFID is useful for an identification in which it uses a tag or small chip with antenna, which carries data and it can be read by the RFID reader. These data travel through radio waves; it is as similar as barcode technology. The main difference between barcode technology and RFID identification is distance. Barcode requires line of sight between users and the tag, but RFID can identify up to hundreds of meters [50]. It decreases the parental figure's heaps in home checking and causes them to screen the patients experiencing incessant ailments. The RFID framework in social insurance comprises two primary segments: radio sign transponder (tag) joined to an article (patient or clinical gadgets) and the pursuer. The label consists of two segments: a chip to store the extraordinary character of the article and a reception apparatus to permit the chip to speak with the peruser utilizing the remote medium. The user creates a radio recurrence field to recognize questions through reflected radio influxes of the tag. RFID works by sending the label's number to the user utilizing radio waves. An ONS (object naming services) looks into the label's subtleties from a database, for example, when and where it was made. RFID tags are the two type's active and passive tags. Active tags have an additional power source, but passive tags have no power source. As per the demand, the user can use these tags.

RFID technologies are again divided into two types: near and far. A near-RFID reader uses a coil, which generates a magnetic field while passing the current. These tags have a small coil and it can produce small magnetic fields that encode the signal to be transmitted. Far-RFID has an antenna in a reader and it can generate electromagnetic waves. This tag also uses a dipole antenna and it can transmit messages by using this power.

RFID technology is very useful in many applications like access control, identification, supply chain management, security, authentication, and tracking. The RFID tag detects the object and stores that information for future use if required. It can also control the movement of objects. Fast tag is also an example of RFID technology. RFID technologies are more secure, easy to install, and easy to use.

2.4.5 Cloud in Healthcare

As per today's demand, cloud computing is a very useful area for storing the collected data. In IoT healthcare, a huge amount of data is generated, so that we need an efficient system to handle this type of data [51]. Cloud computing plays an important role in human healthcare services and applications for making it more effective, speedy, long lasting, and more secure. Cloud computing with IoT can enhance the capacity of storage and efficiency. Cloud computing can directly access the data of IoT, so the security is an essential issue. Cloud services are divided into three categories, according to the usages that are shortly discussed below.

Software as a Service (SaaS): Gives applications to human services suppliers that will empower them to work with well-being information or perform other significant errands. It is an initial service of cloud for IoT healthcare.

Platform as a Service (PaaS): Gives apparatuses to virtualization, organizing, and databasing the executives, and that's just the beginning for making IoT healthcare much easier, comfortable, and secure.

Infrastructure as a Service (IaaS): Gives the physical foundation to capacity, servers, and that's just the beginning. These administrations can be utilized to accomplish an assortment of undertakings, yet two key uses are handily recognized in the writing; huge information for the executives and information handling. These two distinct ideas are introduced independently right now. Nonetheless, it is likewise featured that both are basic for a best-in-class IoT social insurance framework, and along these lines ought to be remembered together for future cloud framework plans. This is the interface between client and hospital team that can connect, interact, and perform some operations.

In IoT healthcare cloud, there are huge amounts of data generated within a minute, so to maintain this cloud is an essential and compulsory step. It includes some functions:

Data storage: To sense and collect the data from different sources is required to store. After storing the data, doctors can monitor this data and further process on it. To analyze the condition and monitor the performance, storage is very important.

Data analysis: This system generates a very huge amount of data, so analysis of data is highly required. Here machine learning and data mining techniques are useful for extracting the useful data from the collected data.

Data cleaning: After data sensing and data analysis, it is very important to clear the useless data. Sometimes, duplicate data consume large amounts of storage and then you can diagnose these data and should clean this for smooth function.

With all these IoT cloud in healthcare focuses toward many important key points; for example, on demand self-services (for providing better availability and accessing), broad network accessing (it is not restricted to the limited space and objects, as the user can access the devices very easily), resource pooling (it means effective use of resources, and it can share the resources with different users as per the use), and handle the rapid changes (these services should be very dynamic so that it can handle the change quickly). In healthcare, the function may change quickly, and in that case, the system must be ready for measurable different functions and handle the rapid changes. Cloud computing plays a very important role in IoT healthcare, or we can say it plays a role of backbone of this system.

2.4.6 Big Data Management

The healthcare industries are already utilizing the benefits of digitization. They maintain their medical records, pharmacy companies, research updates, and many more through electronic database or cloud-based solutions. Big data management is very popular with the five keys that are capacity, speed, assortment, authenticity, and worth. In healthcare applications, managing a huge amount of data with proper planning and function is very important. Big data plays an important role for managing the generated data through different devices.

The term *big data* includes the incredible amount of data flowing over the cloud or digital world. In the healthcare world, the patient and hospital team generates data exponentially because of digital environment of communication. In IoT healthcare, the use of wearable technologies increases the amount of medical data over the cloud. It estimates over 11 billion devices and sensors connected to the internet that will be growing on a very fast rate.

Big data in healthcare is useful for prediction, curing diseases, monitoring the functions of different sensors, avoiding unnecessary death, and improving the life style of patients or normal human lives. Through IoT healthcare, an active engagement is available for day and night, and the consumer always feels safe and sound because he or she is under the supervision of the doctor. Big data is useful to deliver needed outcomes of patients to the system to ensure the security.

The term "connected health" is used to explain how healthcare became the part of the digital health sector. IoT-based healthcare can signify advancement in communication devices, softwares, hardwares, network, data analysis, big data, and cloud computing.

2.4.7 Security

These days, the shrewd frameworks are primarily utilized for different applications. The progressions in the data and correspondence advancements encourage scientists and designers to understand that the framework is validated for all applications. If the truth be told, by considering the secrecy of clinical information, a human services framework must satisfy propelled and get to control systems with exacting security and information quality necessities. Brilliant human services application framework is the most noteworthy applications for the doctor to screen the state of comparing patients remotely through WBAN, in light of IoT social insurance conditions. Nevertheless, for IoT conditions, the hugest trademark is the security and the protection of the framework, yet it is still inquiry-capable in different IoT models.

Novel authentication: The new confirmation and the key concur means convention for data security and privacy. It was critical to making sure about the correspondence channel, clinical sensors or gadgets, and the remote servers. To accomplish those prerequisites, the authors presented the confirmation and the key understanding expert tool, which was a lightweight asset compel sensor, and it is reasonable to secure delicate well-being-related information. In this way, the professionally presented framework gives progressively secure correspondence, ensured the touchy well-being-related information, and the gadgets are modest and little enough in size. Medical information is the most sensitive information, so to maintain the novel authentication is an essential requirement of IoT healthcare.

System privacy: The general framework security and secrecy are by and large interchangeable. The security speaks to the correct control access and it incorporates physical protection. The protection handles the individual in-arrangement through security standards, just as the secrecy obliges the medicinal services doctors to keep their patient's inappropriate individual well-being data. When data are stored in the cloud, then the privacy became a very critical issue, and to maintain this challenge is a very big area of research. As the use of digital frameworks increases, the security flows are increased rapidly. In IoT healthcare system, privacy is a most supervised feature for gaining the trust of clients and makes this system serviceable. The overall function of IoT healthcare is defined in Figure 2.5.

2.5 DIFFERENT CASE STUDIES

In today's world, people became more active and serious toward their health. As we all know, in the fast-growing world no one has much time to take care of them. Nowadays everything is digital, and a huge amount of research also makes healthcare as a digital service. Now people can manage or trace their health through sensors, small devices, or smart mobile phones. E-health has a great initiative in the field of healthcare. It can provide services at the door and has no need to present the physical. It is a combination of medical services, communication services, cloud computing, big data management, and security. It provides medical services for diagnosis,

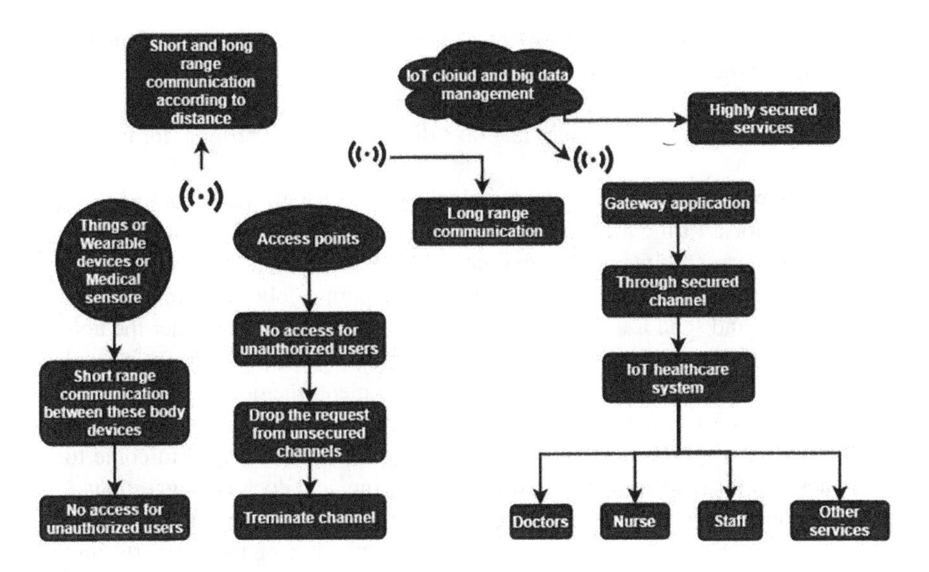

FIGURE 2.5 Function at every layer of IoT healthcare.

monitoring, and controlling without any physical appearance. There are different cases in IoT healthcare, which are discussed further.

Static nodes: In E-health IoT applications, there are many case studies that are present. One of the common studies is a static nodes study. In this type of scenario, sensor nodes are static; for example, patient is on a bed and has no motion or coma condition. In these conditions, patients have no motion or very low motion. In these types of cases, static sensor nodes are sufficient to monitor the patient condition and send the collected results to the access point and then further. Doctors can observe the collected results and if results are normal, then there is no need of any action, and this process is continuing. This case study clarifies that it consumes less energy, it generates less difference, less involvement of hospital team like doctor, nurse, and staff. This case also has less interaction with online services and application. This is useful for monitoring purposes; it may reduce the cost of the overall system.

Dynamic nodes: In IoT-based healthcare, this is a very important case study of patients and doctors. If the patient is partially fit and can move, this kind of application is useful. In this situation, wearable devices or medical sensors are attached with the patient body so that the sensors can move with the patient and sense the data actively. There are many wearable devices according to the requirements, as we already discussed about these devices. These sensors collect the current information of the patient body and send it over the internet and it is received by the hospital. If everything is under normal situations, then there is no need to monitor continuously, but if the situation is not normal, an alarm indicates to the doctor and doctor gives immediate contact to the patient. If the patient condition can be handled through the online suggestion, then the doctor can go for it; otherwise, send a doctor or nurse to the patient home. If the situation is not under control, then send an ambulance and take the patient to the hospital for further treatment.

Normal reports: Wearable devices always connect with the patient and are active for all time and collect the data from the human body and send it to the access points after a few intervals. A cloud data base is very well designed for each and every action. Access points collect the report from sensors and send it to the cloud. If the collected units come under the normal units, then no need to panic; otherwise, an automatic alarm is started and alerts a doctor. According to the requirement, the doctor immediately takes a smart action. It will reduce the time and cost in terms of traditional healthcare system. In IoT healthcare, the patient has no need to come to hospital for regular checkup, health monitoring, and doctors' suggestions.

Abnormal reports: Abnormal reports are the complicated part of IoT healthcare. Collecting the patient data and reports, monitoring these reports, and giving suggestions to the patients; these are the very balanced steps of IoT healthcare. But in case of abnormal reports, it generates some complications. Sometimes patients panic and increase the abnormality unnecessarily. Due to some hardware issues, if the alert alarm is not active, it may generate very serious results and many more. If reports are very slightly different, then the doctor then handles the situation online, but if the differences are very high, then the doctor refers a best doctor to that location or sends an ambulance to take the patient in the hospital.

Secure agreement: In IoT healthcare one of the most important points is security. IoT healthcare generates a very huge amount of data at every second. If this system is not secure, then any unauthentic user can access the data and create misunderstandings, change data, and send false messages to the users. So, the security is a very basic and essential point for IoT healthcare. Before sending over the cloud, if a secure channel is created, then communication takes place, and if the keys do not match, the communication is cancelled.

2.6 FUTURE SCOPE

This chapter is focused toward the IoT, its application, challenges, and many more. For improvement of IoT healthcare, machine learning techniques can play a very important role. This work can be made more appropriate and exact with the help of machine learning techniques. In the future, this study can be done by using supervised or unsupervised techniques.

2.7 CONCLUSION

IoT, WSN, body area sensor network, and smart healthcare are becoming very interesting fields of research in the medical industry. In this chapter, we have discussed the IoT in detail and also discussed the IoT in healthcare. IoT itself includes many features, and we have tried to cover many of the important factors of IoT. This chapter also focuses toward the sensing, wearable devices, communication, and different case studies. The main contribution of this chapter is to compare old and new IoT infrastructure and discuss the latest technologies in healthcare. Nowadays the internet creates their existence everywhere, and very critical areas are dealing with the help of the internet. The services of internet are like anytime, anywhere, and anything. It makes human life smarter and easy as compared to traditional technologies. In the traditional healthcare system, a patient should present at hospital in front of a specific doctor for any health issue. It consumes lot of efforts in terms of time, cost, and resources. To reduce this type of typical services, IoT healthcare is one of the best solutions to provide these services every time and everywhere. The overall purpose of this chapter is to study the infrastructure of the IoT, and its applications and features in the field of healthcare. In this chapter, the authors first discuss the wireless technologies, second the IoT, and then the IoT healthcare-based research in terms of devices, communication technologies, cloud, and security. Afterward, a detailed discussion of different cases generated in healthcare services and future has been given.

REFERENCES

1. A. Prasmitha, "A Review on IoT Devices," International Journal of Advanced Science and Technology, vol. 29, no. 4, pp. 54–62, 2020.
2. Mir Nazish and M. Tariq Banday, "Green Internet of Things: A Study of Technologies, Challenges and Applications," 2018 International Conference on Automation and Computational Engineering (ICACE), IEEE, 2018.
3. Hamza, Khemissa and Djamel Tandjaoui, "A Lightweight Authentication Scheme for E-Health Applications in the Context of IoT." 2015 9th International Conference on Next Generation Mobile Applications, Services and Technologies, 2015. DOI: 10.1109/ngmast.2015.316.
4. M. Maksimović and V. Vujović, Internet of Things Based E-health Systems: Ideas, Expectations and Concerns. In: Khan S., Zomaya A., Abbas A. (eds) Handbook of Large-Scale Distributed Computing in Smart Healthcare (Scalable Computing and Communications). Springer, New York, 2017.
5. Hisham N. Almajed, Ahmad S. Almogren, and Ayman Altameem, "A Resilient Smart Body Sensor Network through Pyramid Interconnection," IEEE Access, vol. 7, pp. 51039–51046, 2019.
6. Cham, Suhardi and Alfian Ramadhan, "A Survey of Security Aspects for IoT in Healthcare," Lecture Notes in Electrical Engineering Information Science and Applications (ICISA), pp. 1237–1247, 2016. DOI: 10.1007/978-981-10-0557-2_117 12.
7. J. Lee, W. Lin, and Y. Huang. A Lightweight Authentication Protocol for Internet of Things. 2014 International Symposium on Next-Generation Electronics (ISNE) 2014. DOI: 10.1109/isne.2014.6839375.
8. S. Zeadally, J. T. Isaac, and Z. Baig, Security Attacks and Solutions in Electronic Health (E-health) Systems" Journal of Medical System, vol. 40, p. 263, 2016. Retrieved from https://link.springer.com/article/10.1007%2Fs10916-016-0597-z (29/09/2018).
9. Md. Mahmud, Hossain, et al. "Towards an Analysis of Security Issues, Challenges, and Open Problems in the IoT." 2015 IEEE World Congress on Services, 2015. DOI: 10.1109/services.2015.12 3.
10. Patricia A H, Williams and Vincent Mccauley, "Always Connected: The Security Challenges of the Healthcare IoT." 2016 IEEE 3rd World Forum on IoT (WF-IoT), 2016. DOI: 10.1109/wf-iot.2016.7845455 2.

11. Subrata Dutta, Mohammad S. Obaidat, Keshav Dahal, Debasis Giri, and Sarmistha Neogy, "Comparative Study of Different Cost Functions Between Neighbors for Optimizing Energy Dissipation in WSN," IEEE System Journal, vol. 13, no. 1, pp. 289–300, 2019.
12. Subrata Dutta, Nandini Mukherjee, Monideepa Roy, and Sarmistha Neogy, "A Study on Efficient Path Selection Algorithms for Propagating Data Messages with a Goal of Optimising Energy Dissipation in WSN," International Journal of Sensor Networks, vol. 15, no. 4, pp. 199–213, 2014.
13. C. C. Y. Poon, et al. "Body Sensor Network: In the Era of Big Data and Beyond," IEEE Reviews in Biomedical Engineering, vol. 8, pp. 4–16, 2015.
14. A. M. Uddin et al. "Internet of Things Technologies for Healthcare," Springer, Fourth International Conference on Internet of Things (IoT) Technologies for HealthCare, 2016.
15. Australian Institute of Health and Welfare, "Australia's Health," 2014. [Online]. Available: http://www.aihw.gov.au/WorkArea/DownloadAsset.aspx-?id=60129548150.
16. E. Perrier, "Positive Disruption: Healthcare, Ageing & Participation in the Age of Technology," Australia: The McKell Institute, 2015.
17. P. Gope and T. Hwang, "BSN-Care: A Secure IoT-Based Modern Healthcare System Using Body Sensor Network," IEEE Sensors Journal, vol. 16, no. 5, pp. 1368–1376, 2016.
18. P. P. Ray, "A Survey on Internet of Things Architectures," Journal of King Saud University, Computer and Information Sciences, vol. 30, no. 3, pp. 291–319, 2016.
19. A Maria, N. Islam, and N. Zaman, "A Lightweight and Secure Authentication Scheme for IoT Based E-Health Applications," IJCSNS International Journal of Computer Science and Network Security, vol. 19, no. 1, January 2019.
20. S. Pallavi and R. Sarangi, "Internet of Things: Architectures, Protocols, and Applications," Journal of Electrical and Computer Engineering, pp. 1–25, 2017. https://doi.org/10.1155/2017/9324035.
21. Boyi Xu, et al. "Ubiquitous Data Accessing Method in IoT-Based Information System for Emergency Medical Services," IEEE Transactions on Industrial Informatics, vol. 10, no. 2, pp. 1551–3203, May 2014.
22. M. Ganesan and N. Sivakumar, "IoT Based Heart Disease Prediction and Diagnosis Model for Healthcare Using Machine Learning Models," 2019 IEEE International Conference on System, Computation, Automation and Networking (ICSCAN), 2019.
23. R. A. Khan and AI. S. K. Pathan, "The State-of-the-Art Wireless Body Area Sensor Networks: A Survey," International Journal of Distributed Sensor Network, vol. 14, no. 4, 2018. DOI: 10.1177/1550147718768994.
24. Sharath Anand and Sudhir K. Routray, "Issues and Challenges in Healthcare Narrowband IoT," 2017 International Conference on Inventive Communication and Computational Technologies (ICICCT), 2017.
25. Stephanie B. Baker, Wei Xiang, and Ian Atkinson, "Internet of Things for Smart Healthcare: Technologies, Challenges, and Opportunities," IEEE Access, vol. 5, pp. 26521–26544, 2017.
26. P. Gope and T. Hwang, "BSN-Care: A Secure IoT-Based Modern Healthcare System Using Body Sensor Network," IEEE Sensors Journal, 2015. DOI: 10.1109/JSEN.2015.2502401.
27. Zaid Alaa Hussien, Hai Jin, Zaid Ameen Abduljabbar, and Mohammed Abdulridha Hussain, et al. "Secure and Efficient e-Health Scheme Based on the Internet of Things," 2016 IEEE International Conference on Signal Processing, Communications and Computing (ICSPCC), 2016.
28. Chaitanya Kulkarni, Snehal Kenjale, Manasi Patil, Chinmay Kulkarni, and S. R Hiray, "Smart Selfregulatory Health System," 2017 2nd IEEE International Conference on Recent Trends in Electronics, Information & Communication Technology (RTEICT), 2017.

29. Naveen, R. K. Sharma, and Anil Ramachandran Nair, "IoT-Based Secure Healthcare Monitoring System," 2019 IEEE International Conference on Electrical, Computer and Communication Technologies (ICECCT), 2019.
30. R. Punj and R. Kumar, "Technological Aspects of WBANs for Health Monitoring: A Comprehensive Review," Springer Science + Business Media, LLC, part of Springer Nature, Wireless Networks, 25, 1125–1157, 2019. https://doi.org/10.1007/s11276-018-1694-3.
31. Hesham A. El Zouka, "An Authentication Scheme for Wireless Healthcare Monitoring Sensor Network," 2017 14th International Conference on Smart Cities: Improving Quality of Life Using ICT & IoT (HONET-ICT), 2017.
32. S. M. Riazul Islam, et al. IEEE Access "The Internet of Things for Health Care: A Comprehensive Survey," June 1, 2015, date of current version June 4, 2015. 10.1109/ACCESS.2015.2437951.
33. Sahmi Imane, Mazri Tomader, and Hmina Nabil, "Comparison Between CoAP and MQTT in Smart Healthcare and Some Threats," 2018 International Symposium on Advanced Electrical and Communication Technologies (ISAECT), 2018.
34. N. Zhu, T. Diethe, M. Camplani, L. Tao, A. Burrows, N. Twomey, D. Kaleshi, M. Mirmehdi, P. Flach, and I. Craddock, "Bridging e-Health and the Internet of Things: The SPHERE Project," IEEE Intelligent Systems, vol. 30, no. 4, pp. 39–46, 2015.
35. S. H. Chang, R. D. Chiang, S. J. Wu, and W. T. Chang, "A Context-Aware, Interactive M-Health System for Diabetics," IT Professional, vol. 18, no. 3, pp. 14–22, 2016.
36. Maria Almulhim and Noor Zaman, "Proposing Secure and Lightweight Authentication Scheme For IoT Based E-Health Applications," 2018 20th International Conference on Advanced Communication Technology (ICACT), 2018.
37. Jinjin Zheng, Xia Hu, and Hang Fang, "The Application of Labview in Mine Hydrology Wireless Monitoring System," 2014 IEEE Workshop on Electronics Computer and Applications, 2014.
38. C. F. Pasluosta, H. Gassner, J. Winkler, J. Klucken, and B. M. Eskofier, "An Emerging Era in the Management of Parkinson's Disease: Wearable Technologies and the Internet of Things," IEEE Journal of Biomedical and Health Informatics, vol. 19, no. 6, pp. 1873–1881, 2015.
39. Y. J. Fan, Y. H. Yin, L. D. Xu, Y. Zeng, and F. Wu, "IoT-based Smart Rehabilitation System," IEEE Transactions on Industrial Informatics, vol. 10, no. 2, pp. 1568–1577, 2014.
40. C. Doukas and I. Maglogiannis, "Bringing IoT and Cloud Computing Towards Pervasive Healthcare," Proceedings of Sixth International Conference on Innovative Mobile and Internet Services in Ubiquitous Computing (IMIS), 2012, pp. 922–926.
41. S. Sarkar and S. Misra, "From Micro to Nano: The Evolution of Wireless Sensor-Based Health Care," IEEE Pulse, vol. 7, no. **1**, pp. 21–25, 2016.
42. Y. Yin, Y. Zeng, X. Chen, and Y. Fan, "The Internet of Things in Healthcare: An Overview," Journal of Industrial Information Integration, vol. 1, pp. 3–13, March 2016.
43. Shahnaz, Saleem, Sana Ullah, and Kyung Sup Kwak, "A Study of IEEE 802.15.4 Security Framework for Wireless Body Area Networks," Sensors, vol. 11, no. 2, pp. 1383–1395, 2011.
44. Muneer Bani Yassein, Wail Mardini, and Amnah Al-Abdi, "Chapter 9, " Security Issues in the Internet of Things," Critical Research on Scalability and Security Issues in Virtual Cloud Environments, IGI Global 2018, pp. 186–200.
45. Noha M M. AbdElnapi, Nahla F. Omran, Abdelmageid A. Ali, and Fatma A. Omara, "A Survey of Internet of Things Technologies and Projects for Healthcare Services," 2018 International Conference on Innovative Trends in Computer Engineering (ITCE), 2018.

46. Mrinai M. Dhanvijay and Shailaja C. Patil, "Internet of Things: A Survey of Enabling Technologies in Healthcare and its Applications," Computer Networks, vol. 153, pp. 113–131, 2019.
47. Shashank P. Wankhade and A. N. Bandal, "Security for Automation in Internet of Things Using One Time Password," 2017 International Conference on Computing, Communication, Control and Automation (ICCUBEA), 2017.
48. Dalal Abdulmohsin Hammood, Hasliza A Rahim, Ahmed Alkhayyat, and R. Badlishah Ahmed, Qammer H. Abbasi, "Reliable Emergency Data Transmission Using Transmission Mode Selection in Wireless Body Area Network," Cogent Engineering, vol. 5, no. 1, 2018.
49. Veeraselvam Aruna, Mohammed Gulam Nabi Alsath, Savarimuthu Kirubaveni, and Marimuthu Maheswari, "Flexible and Beam Steerable Planar UWB Quasi-Yagi Antenna for WBAN," IETE Journal of Research, pp. 1–11, 2019.
50. G. Kumaresan and N. P. Gopalan, "Programmable Cellular Automata Based Random Key Generation for One-Time Pad Encryption Using Unity Attractors," Journal of Applied Security Research, pp. 1–18, 2019.
51. Ali, Rahat and Al-Sakib Pathan, "The State-of-the-Art Wireless Body Area Sensor Networks: A Survey," International Journal of Distributed Sensor Networks, vol. 14, no. 4, 2018. http://doi.org/10.1177/155014771876899.

3 Universal IoT Framework

Joy Dutta and Sarbani Roy
Jadavpur University, West Bengal, India

CONTENTS

3.1 Introduction ... 45
3.2 Requirement of the Framework ... 48
3.3 Related Work ... 48
3.4 IoT Ecosystem ... 49
3.4.1 Connectivity Technology ... 49
3.4.2 Messaging Technology ... 51
3.4.3 Platform Technology ... 51
3.5 Universal Framework for IoT Applications ... 52
3.5.1 Data Layer ... 54
3.5.1.1 Government-Related IoT Platforms ... 54
3.5.1.2 Enterprise/Company-Based IoT Platforms ... 54
3.5.1.3 Business-Oriented IoT Platforms ... 55
3.5.2 IoT Gateway: Edge/Fog Layer ... 55
3.5.3 Cloud Layer ... 55
3.5.3.1 IoT Middleware ... 56
3.5.3.2 IoT Service Layer ... 59
3.6 IoT-Based Real-World Applications and Unified Framework ... 60
3.7 Mapping of Big Data Framework for IoT to Universal Framework for Potential Applications ... 62
3.8 Challenges ... 64
3.8.1 Security ... 65
3.8.2 Interoperability ... 66
3.8.3 Governance ... 66
3.9 Summary ... 66
References ... 67

3.1 INTRODUCTION

In the current days, Internet of Things (IoT) is a buzzword used smartly to address every modern-day problem. Today, it is commonly used in several fields, and approaches are recognized worldwide because of its intrinsic intelligent design. Taking into account all sorts of sensors installed in devices and infrastructures, such as homes, cars, highways, traffic lights, power grids, greenhouses, and the surrounding environment, there is a possibility for more than 50 billion devices to be associated by 2030 [1]. The growing use of IoT in different fields is also posing new challenges in terms of interoperability. Different fields have different criteria as well as different specifications from the state-of-the-art solution provider. Due to this difference in standards, various frameworks are in place in different domains. Therefore, though IoT is a popular solution point, because of the variety in the application domain, a standard framework for IoT is a requirement in today's perspective.

Architecting a new device to be interoperable with other devices already in the market, or infrastructures already in place in the system, has grave significance in the field of IoT, particularly because there is a vast number of diverse gadgets associated with big IoT platforms. Online approval sets permit gadget traders in checking the items in interoperating with a vast assortment of different trader items. Checking for interoperability with a legitimate product that has not made any progress or is yet to be available for ads is impractical. Nevertheless, online approval suites allow interoperability testing with standard protocols and frameworks, ensuring compatibility with peer IoT gadgets that have not yet made progress. Nonetheless, there are likely to be interoperability holes. For example, data models developed by competing standards that have syntactic differences, while semantics may be identical. Thus, generic protocols cannot be completely interoperable if qualification testing is incomplete or not thorough. Hence, simulation tools that essentially deploy client-specific configurations would be beneficial. Simulations help to identify interoperability differences in requirements and testing frameworks for program behavior and data definitions. Trial implementations and testbeds are a common method for identifying these gaps. It helps to define hardware-based incompatibilities. So, deployments for the trial must go live before the gaps are to be resolved. Testbeds should be used for longer-term research and development with the concurrent implementation of enhancing functionality and features, while ensuring that interoperability or backward compatibility problems are not addressed.

There are typically two types of architectures in IoT that are suitable for the implementation of all IoT-based applications. Those are service-oriented architecture (SOA) and API-oriented architecture [2]. SOA ensures interoperability between heterogeneous devices. In generic SOA, we generally have four layers, with distinguished functionalities. At first, a sensing layer that is combined with the hardware artifacts available to detect the status of the actual deployed devices. Secondly, the network layer, which is the infrastructure to enable connection through either wireless or wired links between the devices. Thirdly, the service layer consists of designing and managing the services provided by users or applications. Finally, the interface layer consists of the methods of interacting with users or applications. Throughout this type of design, complex system structures are divided into subsystems that are loosely coupled and can be reused later, thereby offering a simple way to manage the whole complex system by taking care of its individual subsystems [3]. This will ensure that the rest of the system can still function normally in the event of a component failure. This is of tremendous importance for the efficient design of the IoT application architecture, where reliability is the most critical parameter. If SOA can bring these facilities to IoT architecture, then features like interoperability and scalability can be increased for IoT entities. Also, from the user's point of view, all resources are abstracted, and thus the user's burden is reduced to deal with various layers and protocols. Besides, SOA has the ability to create diverse and complex systems by using modular composability, where the execution of each task involves a sequence of service calls to all the different modules, which may be distributed over several locations.

Conventional approaches to designing service-oriented solutions use SOAP and Remote Method Invocation (RMI) as a way of defining, finding, and calling services. However, due to the overhead and difficulty of these approaches, Web APIs and REST methods have been implemented as favorable substitutes. The needed resources for these substitutes vary from network bandwidth to computational and storage capacity and are generated by request-response data transformations that occur periodically through service calls. Lightweight data-exchange formats like JavaScript Object Notation (JSON) can decrease the aforementioned overhead, particularly for smart devices and sensors with finite resources, by eliminating bulky XML files used to represent facilities. This helps to allow more effective use of the contact channel and to harness the strength of the apps. Similarly, creating different IoT APIs lets the service provider interest a large number of consumers by concentrating more on product quality instead of product presentation. In addition, the security features of modern Web APIs (e.g., Oauth) are also capable of improving the organization's service perceptibility and advancement, which allows multi-tenancy. This also offers more effective control and pricing methods for utilities than previous service-oriented strategies. The API Economy is also concerned with developing utilities and making them accessible in an open manner, all of which drive digital transformation. The APIs can be used by both types of applications, i.e., external as well as internal ones, as they are more handy and unambiguous. These APIs are generally self-sustainable with very little or no guidance, which makes them well-matched for bulk use. While APIs are primarily associated with REST and JSON, other practices, such as web sockets, MQTT, etc., are gaining popularity for different use-cases.

SOA is more than just a protocol. It is an architectural best practice for constructing decoupled applications and promotes re-use of services. They are primarily used for internal use-cases, although they are prevalent in many external B2B scenarios. The architectural concepts of security, compliance, policy management, monitoring, and analysis of SOA are the same or similar to API. Therefore, while APIs have their unique characteristics, at the heart they are not that different from SOA, and organizations should aim and use a universal infrastructure and manage and control cohesion between them, rather than implementing a redundant infrastructure. So the architecture based on SOA and API has different advantages. If one thinks of IoT architecture that serves all existing applications, it is not possible to accommodate all of them using a single one.

Hence, a hybrid of these two can be a solution for the universal framework requirement. In this way, comprehensive framework engineering for IoT needs to ensure the perfect activity of its parts (dependability is considered as the most import configuration factor in IoT) and connection of the physical and virtual domains together. To accomplish this, a cautious thought is required in planning failure recovery and versatility. Furthermore, since versatility and a dynamic difference in the area have become a necessary part of IoT frameworks with widespread utilization of cell phones, cutting-edge structures need to have a specific degree of flexibility to appropriately deal with dynamic collaborations inside the entire ecosystem.

3.2 REQUIREMENT OF THE FRAMEWORK

IoT architectures have four key design objectives: (1) lessen production period and get IoT technologies to market more quickly; (2) decrease the apparent complexity of IoT implementation and operation; (3) improve portability and interoperability of applications; and (4) improve modifiability, reliability, and maintenance. Considering the broad variety of current and evolving communications technology choices, it is impractical for applications to handle combinations of alternative ways to communicate. Frameworks cover the complexities of networking under a higher-level message passing abstraction like REST and publish-subscribe. Standards and organizations are helping to realize these objectives by standardizing the network layer interconnect, the message passing interface description along with the application-led description of data.

IoT Frameworks try to simplify its internal networks by providing an abstraction that hides most of the underlying complexity, although revealing data, interfaces, and functions that make it easier to interoperate. To build an IoT-based application, high-level languages are used, which hide the system's internal complexity by utilizing the framework. The framework gives space to IoT network designers to focus on node interaction semantics rather than connectivity information, as it takes care of the rest of the things. This also enables increased portability of different IoT-based applications. Portability can be done on various levels. In the case of a bottom layer of the framework, the framework is generally exclusive to the operating system, whereas the top layer of the framework is the case-specific situation. IoT applications can be built based on a software abstraction and can run on any OS to which the framework is being ported. The framework also enables interoperable devices in heterogeneous environments and builds connectivity intelligence in it.

3.3 RELATED WORK

The surplus and proliferation of devices in IoT is a new challenge for the architectures that control the interaction and communication between smart objects. Numerous research attempts have been made to fulfill these new standards. These works are grouped into three categories: (1) IoT infrastructure, (2) network infrastructure, and (3) cloud infrastructure.

The IoT Infrastructure Working Group recommended direct changes to the physical infrastructure that would allow interaction between smart objects [4]. In particular, these works suggested strategies for topology control to provide fault tolerance by smart device placement or through the use of the communication network to provide alternate and simultaneous routing paths. A set of mechanisms focusing on a global view of the network infrastructure to enable smart decisions is proposed in [5, 6] from the IoT infrastructure to the cloud and cloud environments. The Software-Defined Networking (SDN) methodology is used in the research referred to in [7] to increase the granularity of the network infrastructure and to improve the efficiency of the network. Works in the cloud infrastructure category [8, 9] tackle the resource limitations of IoT applications, taking advantage of the cloud model. Such researchers' proposal deals with the usage of resource-conscious cloud environments as well as the reliability of services. Taking into account the features and criteria of the IoT infrastructure that support smart cities reinforces the need for a fresh IoT communication architecture. The lack of support from the communication infrastructure for IoT services calls for a framework that requires the heterogeneity, adaptability, and resilience needed in this type of dynamic scenario.

3.4 IoT ECOSYSTEM

Before going towards the unified framework for IoT, understanding its ecosystem is the crucial first step that needs to be considered. The IoT ecosystem is highly complex, fragmented, and growing. The IoT ecosystem is evolving and it is evolving fast. One of the major factors is the device substitution rate, which is different for different devices. One cannot compare the replacement period of a PC in a company with a personal smartphone or with a building's HVAC system. From short to long replacement rates, both can speed up as well as can slow down the adaption of new technologies in different sectors. Due to the many variations in the different sectors of the IoT ecosystem, the sectors tend to accept technology differently, which is reflected in Figure 3.1.

The market forces that sustain the sectors are identified partly due to technological requirements that are specific to the uses and applications that drive the stability of the internal market. Though till now many brownfield solutions have benefited from different exclusive or vertically integrated solutions, long replacement cycles, and costly particular hardware components, this is not going to continue, as innovative IoT technologies are spilling over these, which is a positive thing. Nevertheless, these disruptive forces that break down corporate silos often introduce new challenges to security in the form of increased sophistication, new business models, and unanticipated encounters.

The IoT ecosystem can be described graphically, as shown in Figure 3.2. Technologies within a particular sector of the ecosystem are specific for that particular sector's environment, i.e., its required networking, protocols, platforms and data analysis set up are generally a fixed specific set. For any IoT-based sector, the business designs, as well as production and consumption on that market, are also stable beforehand. In real life, ecosystem-specific components are specialized in different areas of applications. To support this, IoT device elements can be distributed based on performance criteria arising from advanced computing. A key unifying force in all this is the interoperable, low-cost networking capability that makes distributed IoT possible. But addressing various needs across several IoT segments using a single IoT platform seems difficult, but not impossible. Different core areas of the ecosystem are discussed in detail below.

3.4.1 Connectivity Technology

Network and connectivity are of utmost importance for any ecosystem. For creating a sustainable IoT ecosystem system, IoT must be able to connect over all types of distances, e.g., short, medium, and long. It not only requires the connectivity, but along with this, it must follow some transmission quality criterion (e.g., low-latency,

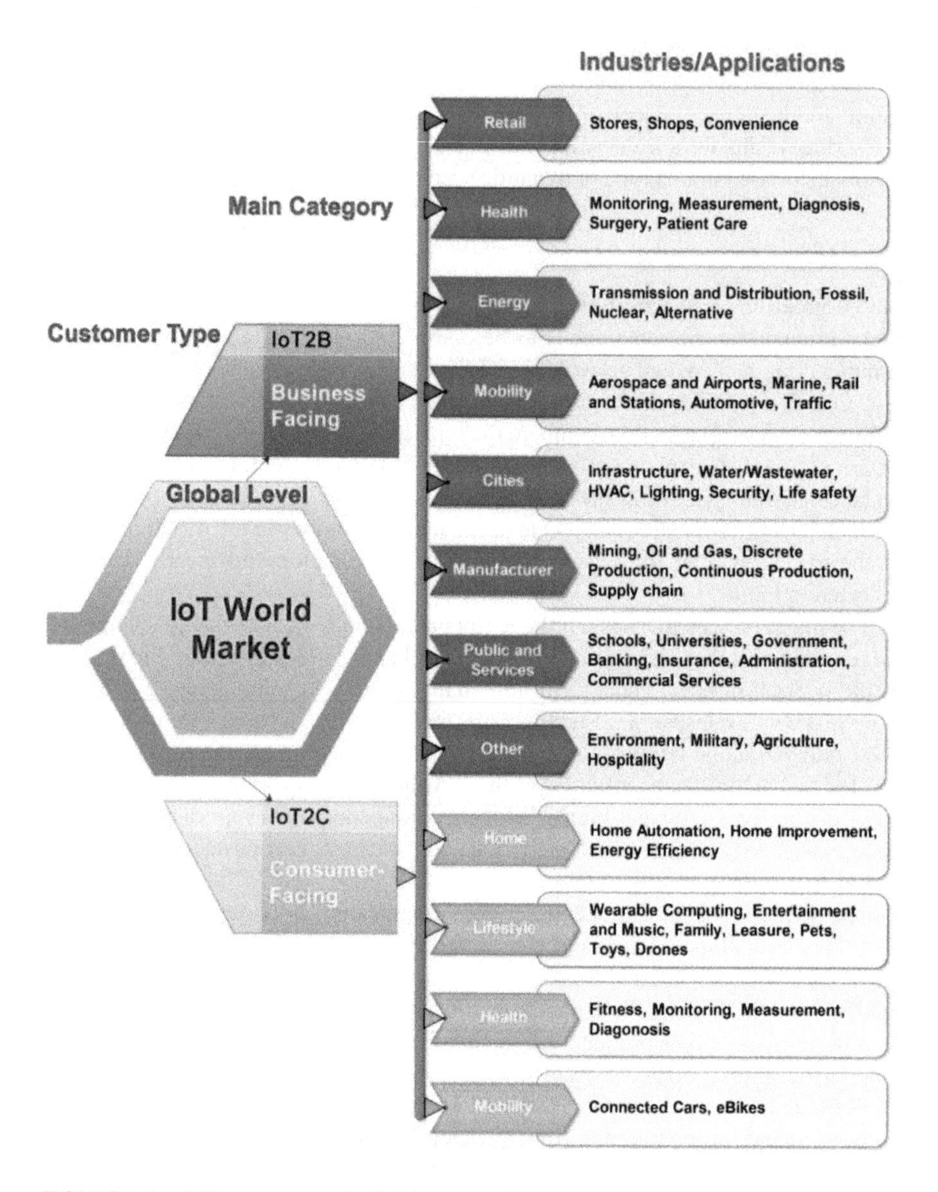

FIGURE 3.1 Different sectors for IoT-based applications.

streaming optimizations, etc.) without which creating a connection between objects are worthless. IoT systems should recognize the presence of other environmental threats which affect the QoS, such as radio interference or emissions from other electronic devices, low power levels, noise, and resource depletion scenarios. Here, guaranteed service level along with quality is the requirement of the state. A variety of network technologies have appeared to address the multi-faceted requirements of IoT, such as ZigBee, LoRa, LPWAN, TSN, etc., whereas some others are specialized

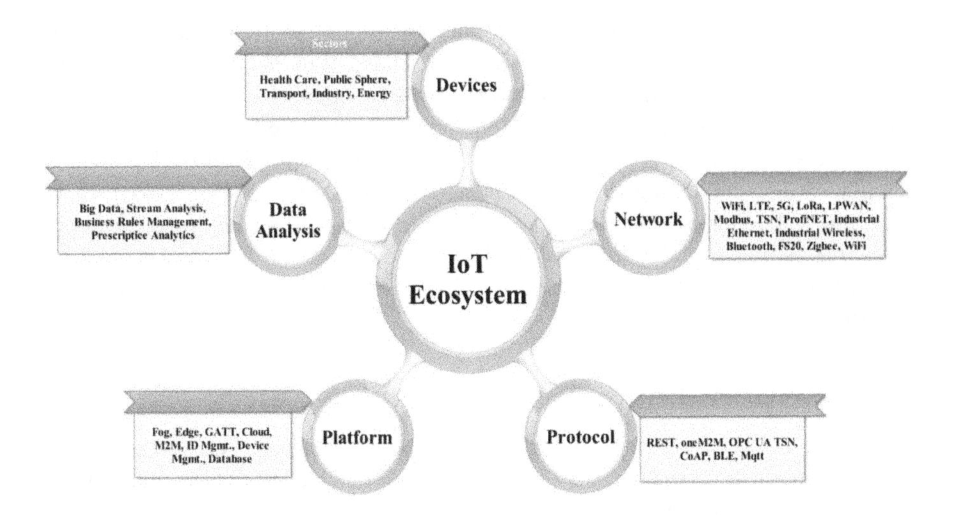

FIGURE 3.2 IoT ecosystem.

in a particular application sense, such as CAN, Modbus, etc. Some serve a more general purpose, e.g., WiFi, Bluetooth, 3G, 4G, 5G, and Ethernet, which includes communication networks, streaming media, and network control systems.

3.4.2 Messaging Technology

Messaging technology defines how messages flow between the nodes of the network. Generally, REST-based approaches are used for messaging purposes, which works on a request-response basis. In different IoT ecosystems, different IoT nodes collect data from other nodes by using different protocols (e.g., HTTP, CoAP, etc.) suitable for that specific application. For this, nodes added extra complexity in the messaging layer for collecting the data for these nodes.

There are certain cases where the publish-subscribe model fits well in a specific application or application scenario. A use case of this model is multicasting or broadcasting, i.e., a certain change of state of a particular node's state can be published to all the neighbors that have subscribed to the node or all the rest of the nodes in that zone. DDS, AMQP, MQTT, and XMPP are examples of these public-subscribe messaging protocols, which are used in different IoT applications based on specific contexts. When different applications are designed for different IoT devices, the connectivity stack needs to be addressed properly for the physical layer of the network which handles these application messages. This complexity is most apparent when an IoT platform supports multiple ecosystems.

3.4.3 Platform Technology

IoT platforms are very essential in terms of successful execution of a user's query for any specific service. There are different platforms available in the IoT framework that supports the realization of an IoT-based application. Fog, Edge, and

Cloud, etc., are different platforms that have specific utilization. These platforms host apps, resources, and data that are useful for the IoT application. These specific platforms are skilled in doing a different type of functionality that each one does. Thus, each has its use cases. Hence, a specific platform is used for a particular purpose or a set of purposes only. These platforms can be optimized for bandwidth, latency, system unloading, bridging through control domains, etc. Among these platform technologies, the cloud holds a very vital role. It is used for a variety of roles, e.g., storage, scalability, and analytics, without which IoT is half-finished. These roles, in turn, help the platform to achieve desired functionality like interoperability and compatibility.

3.5 UNIVERSAL FRAMEWORK FOR IoT APPLICATIONS

The unified framework for IoT, as seen in Figure 3.3, is a descendant of the architecture of the client-server. The Slicing principle has many uses in software engineering, one of which is the re-use of architecture [10]. For this work, architectural slicing is used to demonstrate knowledge of the high-level design of the various IoT-based systems, rather than specifics of the low-level implementation. The proposed architecture is a generalized concept that includes potentially different application activities based on IoT. Here, the different data sources of this framework are connected to the internet directly or smartphone or through the fog and send data to the cloud server. Applications analyze this data in the cloud and deliver a service. This software is capable of mapping not just all smart city-based systems, but also all other IoT-based applications [11]. Each of these applications can be broadly categorized as three types, i.e., IoT-based applications, smartphone-based applications, and a combination of both. Here, smartphone-based applications are treated as a separate category as a smartphone has huge potential in the IoT-based application domain. It can act as a separate IoT device if used in that sense. It can be used as a gadget to get an IoT-based service, or may be used for data forwarding from other IoT devices or can even manipulate IoT-based services. Hence, roles are dynamic and it changes its role from one application to another. This unified framework's mapping to these IoT-based real-life applications (mentioned three types) mapping is discussed later in Section 3.6 of this chapter.

In this architecture, the crowd is one of the most important aspects of the data layer, as until people are smart, all other technologies are useless. So, not only the IoT but also citizens, sensors, smart devices, smartphones, etc., collectively are the main data generating sources of this layer (as seen in Figure 3.3). This layer generates data and is known as a data layer. This layer not only produces data from various IoT-based applications, but it is also responsible for receiving the services offered by these applications. This generalized architecture can be fine-grained further to visualize its internal elements further, according to the requirements of the different applications. The typical aspect with any IoT-based application is the use of various sensors in different applications that are either sensing the environment or monitoring different systems. Then, generally these sensors are transmitting the sensed data to the fog gateway, either for local computation or just for

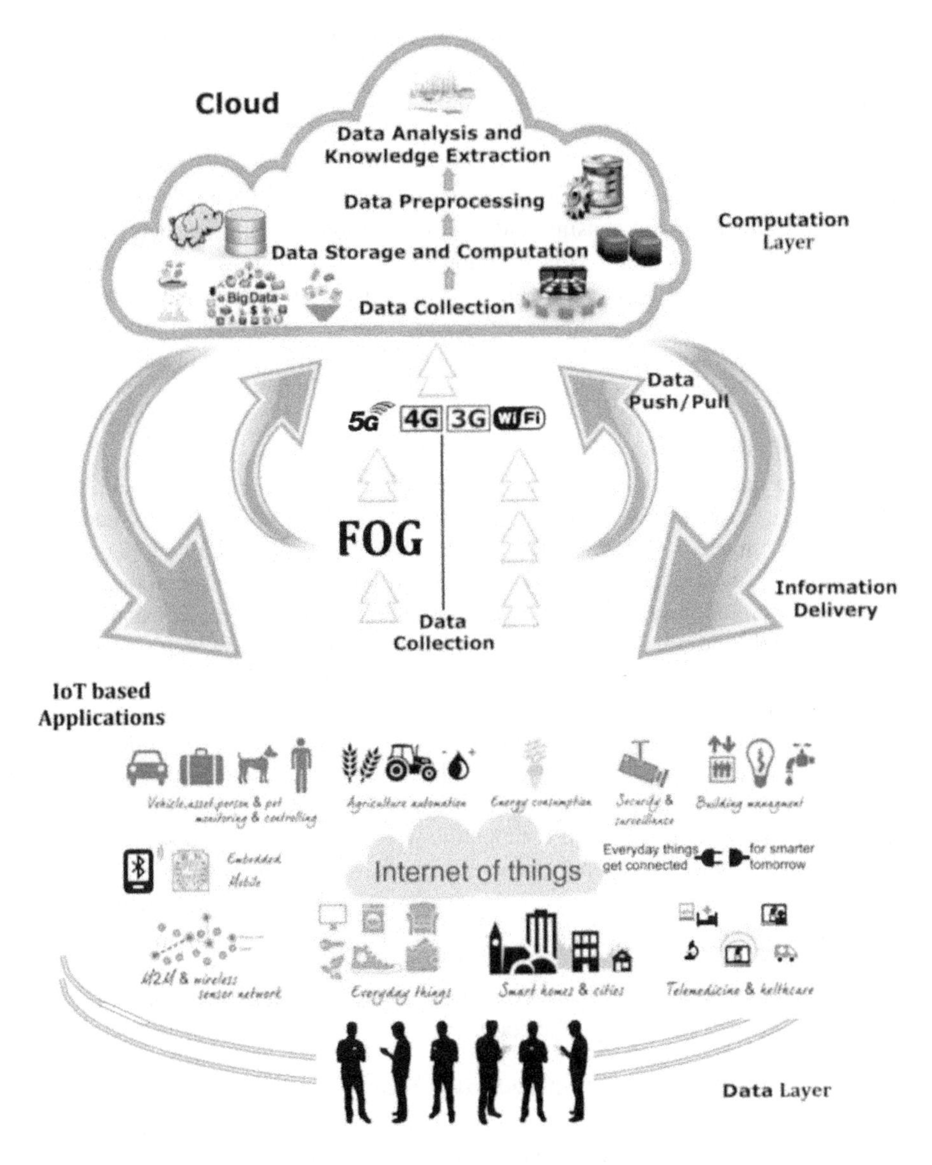

FIGURE 3.3 Unified architecture for IoT-based applications.

forwarding, which is application-specific. The fog gateway has a link to the internet as well, as it has computational and storage capability also. Generally, some of the processing is performed in the fog to experience real-time application behavior. This nature of Fog also allows the system to minimize internet traffic because it helps to determine which data are redundant to send in the cloud layer for further processing. Data collection and power control are the two main roles of fog in this architecture. After the fog layer, the required data are forwarded to the cloud where further data computation and analysis take place. The role of each layer is described below in detail.

3.5.1 Data Layer

The data layer is the base layer which is considered as Tier 1 here. Here, physical objects are attached with sensing and communication capabilities such that they become usable IoT devices. This tier can have several platforms for generating data. Broadly, these IoT platforms may be categorized into three types, i.e., government-based, enterprise/company-based, and business-oriented platforms [12]. This grouping for smoothing the development of IoT is represented in the following subsections.

3.5.1.1 Government-Related IoT Platforms

Government has a large sector in their hand, which is becoming smarter with time. eGovernance is the new key term for government digitization. For the advancement of the government sectors which are related to different smart city services, e.g., environment protection, traffic management, road safety, water conservation, health, education, etc., the government can use the management intelligence using proper IoT framework relevant to each field. Eventually, eGovernment will promote economic growth and governance of the region. It is financed by different government's public welfare program, and this slowly introduces IoT in different sectors of the system. However, this eGovernment framework should not have an "information island" effect, as different sectors are involved in its realization. This framework is required information relevance to avoid this island effect, as well as to ensure application security [13]. Due to their public-welfare properties and social values, this type of IoT platform cannot become a standard business model for a generic IoT platform. Business development guidelines specify that these IoT systems do not have the appeal of business-oriented investment and thus cannot draw venture capital. Thus, these cannot be funded by a third party or business organizations having a business-oriented mindset. They can only be realized separately by the government/public administration for the public good.

3.5.1.2 Enterprise/Company-Based IoT Platforms

To enhance their productivity and quality assurance, market-oriented businesses and organizations need independently funded IoT initiatives. This enhancement requires up-gradation and automation of various sectors of the whole business processes. For this more productive manufacturing, warehousing, distribution, transport, logistics, marketing, and supply chain management is required, which effectively enhances the whole system. Different enterprises or companies use IoT for their internal management. They use different IoT frameworks for the realization of the initiatives taken by different organizations. These initiatives are private and thus required individual private investments for different companies. However, the end products are for the public, and IoT systems produced by these companies penetrate deep into society. Actually, their end product reaches millions of households.

3.5.1.3 Business-Oriented IoT Platforms

Since IoT is the emerging technology and one of the most important catalysts in the industry 4.0 revolution, hence investment in IoT-based systems is very attractive. More investment in the IoT sector helps the growth of the modern economy and accelerates economic growth. This, in turn, helps to build a powerful business model with the successful participation of a large number of end users as well as IoT equipment. For example, the use of radio-frequency identification (RFID) in product logistics includes the automated procurement of logistic data, and logo identification, as well as the accurate distribution of cargo. This also enables secure customs and tracking visualization, which is a much-awaited feature for the customers. Here, the IoT Integrated Information Center is anticipated as an essential part of the IoT Generic Framework. This center is responsible for attracting commercial capital investment in IoT Integrated Application Industry. The information center is run by an IoT service provider, which in addition to the current service-provider categories, can quickly become another major player. The presence and validation of a successful IoT business model would encourage investment and further production of suitable generic IoT platforms.

3.5.2 IoT Gateway: Edge/Fog Layer

The fog gateway server is maintained in Tier 2 of the unified framework, which can make some local decisions based on the data collected by the sensors. This layer is primarily responsible for temporary data storing, filtering, and pre-processing. It also acts as a gateway that links IoT to the internet. In the architecture, the fog node/edge server supports sensor data and rapid reaction in local computing using actuators [14]. This saves cloud connectivity bandwidth and decongests the core network. Actual devices are virtualized here, in this layer, and this creates virtual IoT devices (VIDs) to manage the actual physical layer here, in Fog. This VID is deployed as microelements for this purpose, having the Thing Description, which is very effective for the later stage data handling purposes. The provided VID includes a communication manager for the sharing of data with actual IoT devices. There are different microservices associated with it, which includes important services like data validation, metadata annotation, local data processing, local actuation, etc.

3.5.3 Cloud Layer

The cloud layer is primarily responsible for data storage and computation due to the very nature of the cloud. In the unified framework, Tier 3 is the cloud layer. From the fog layer data are forwarded to the cloud layer [10]. As shown in Figure 3.3, the key role of this layer is shown as an overview basis. This figure can be further zoomed in for better comprehension. Now, here we're going to discuss those inner details. This cloud layer's functionalities can be divided into two parts, i.e., IoT middleware and IoT services for the actual unified framework realization. Such IoT services can also be seen as microservices. Both these sublayers have different purposes, and their utilization is often different from the design as well as the user perspective. Here, an overall database for the cloud server is maintained, which is used to store sensor data. This is in turn used to control the IoT devices or trigger some action based on the user's requirements. These two sublayers are explained below, along with their significance for the unified framework.

3.5.3.1 IoT Middleware

IoT middleware is the internal first layer in the cloud, which handles some very important aspects of the IoT, e.g., heterogeneity, device management, etc. In the data generation layer, plenty of heterogeneous IoT devices are present and data from all the devices are important. Hence, the seamless integration of devices and data is essential to collect those data in a single platform. This IoT middleware layer resolves the need of the state. This layer involves common features and abstraction mechanisms that enclose information of the IoT infrastructure to users and developers to promote interaction between these actors [15]. Research projects that are now available as a product, such as LinkSmart [16] and OpenIoT [17], have made significant contributions to this integration layer; however, there is still scope for significant improvement in the resilience mechanisms needed in the infrastructure. Keeping this in mind, we're using an IoT middleware layer that focuses on improving IoT resilience. Next, a description of the modules as shown in Figure 3.4, of this layer, is discussed.

3.5.3.1.1 Heterogeneity Manager

Because of the availability of heterogeneous devices in the IoT framework, a dedicated heterogeneity manager is required. It provides a common semantic to achieve flawless communication between the IoT data layer and the upper cloud layer. The heterogeneity manager acts as an intermediary between different IoT components located in the cloud and different IoT islands present in the data layer. The main

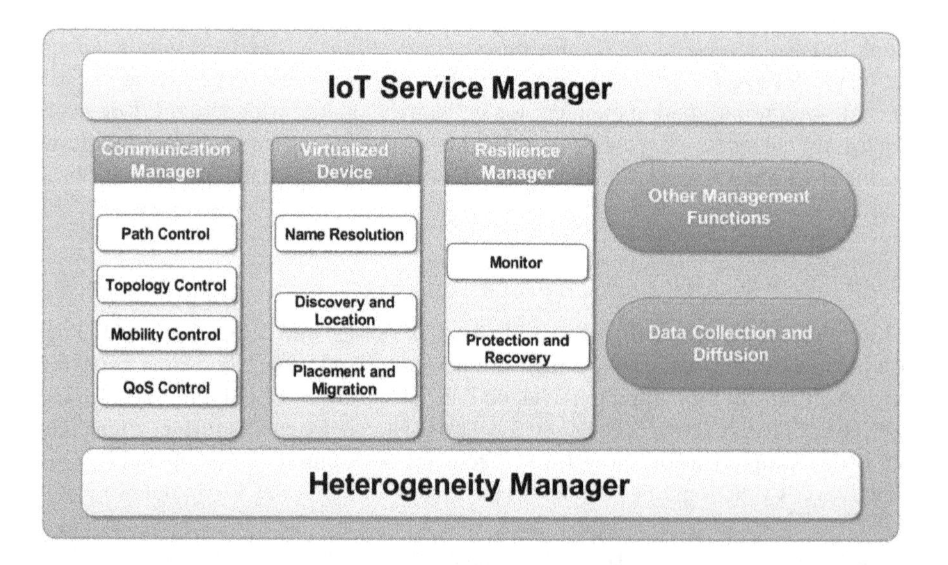

FIGURE 3.4 IoT middleware functionalities in the cloud layer.

function of this part is to convert the communication protocols used in the IoT island into a common language. In this context, potentially strong algorithms for virtual network embedding [18] tackle the convergence of the network and cloud resources, which is a key feature of this IoT framework.

3.5.3.1.2 Communication Manager

The role of the communication manager in the IoT middleware is to provide an effective and adaptive way to manage data travel path, communication network configuration, and mobility of the IoT. It generally controls data sharing, not only between things but also between applications. The communication manager incorporates four sub-modules in general out of which Path Control deals with the complexities of the IoT system and allows mechanisms to adjust the path of particular flows in real-time. Topology Control offers a global view of the infrastructure through knowledge of the status of devices. It also supports these device's communication processes. This module can create multiple virtual networks on the same physical infrastructure, thus enabling an overlayed network topology. The third sub-module in the communication manager module is Mobility Control, which is accountable for storing data in transient buffers on intermediate devices. These buffers hold the data until the service is completed or the entity reaches its destination. Finally, the fourth sub-module is QoS control. This sub-module ensures that the different service's traffic needs are satisfied. This module helps different services to handle different network traffic such that service level agreements are not violated.

To perform the above-mentioned tasks, a mixture of SDN and virtual network (VN) approaches are used to achieve a virtual software-defined network (VSDN) [15]. By decoupling control and forwarding functionality within the communication manager portion, it is possible to integrate required features into the infrastructure that support different IoT-based service and applications, such as dynamism, mobility, and resilience. Concerning infrastructure stability, the prospect of combining different VNs with a multi-path approach using VSDN is a promising approach.

3.5.3.1.3 Virtualized Device Manager

As the name suggests, this module is solely responsible for device management. This component provides ways to define, discover, and locate the resources in the IoT infrastructures. This also enables different devices to move through domains. This module has three sub-modules, as shown in Figure 3.4, which handle this module's functionalities. Functionalities of these sub-modules are mentioned below.

The name resolution sub-module takes into account two basic tasks. The first task is name mapping, i.e., a local name is converted into a flat or agnostic name. The second task is to apply a name resolution strategy to identify each device uniquely. The second sub-module is discovery and location, which monitors the devices and artefacts used in the IoT infrastructure. The last sub-module is the placement and migration module, which is responsible for implementing different IoT device's optimum placement strategy, as well as needing wise virtual device migration that allows instant services or devices to boost or sustain the requirements of different IoT applications while reducing resource consumption. Practically, a migration strategy makes it possible to keep essential services alive even in a faulty scenario.

Some useful methods are defined to perform module tasks within the virtualized system manager. Through network function virtualization (NFV) strategy, it would be possible to virtualize on-demand network services and devices [19]. This strategy can be combined with cloudlets to make the service delivery more versatile. Such technologies allow reducing the consumption of resources in the IoT backbone. This, in turn, allows a higher number of devices to be handled. In addition to NFV, there are other strategies too. Adoption of smart and cognitive technologies to enhance service placement and migration, as well as machine learning strategies, can bring major benefits to the IoT framework by incorporating self-management capabilities [20].

3.5.3.1.4 Resilience Manager

The primary objective of the Resilience Manager module is to provide flexibility to the IoT infrastructure through continuous supervision of the activities. This module has two sub-modules, one of which is Monitor and the second one is Protection and Recovery. The Monitoring module keeps tracking on the actions of the IoT artefacts and triggers other modules to necessary steps when required. The Protection and Recovery module works with route and topology control, as well as placement and migration modules, to ensure that proper action is taken in the event of a failure. It is also worth mentioning that the Resilience Manager is not solely responsible for the overall structure's flexibility, but acts as an orchestrator of all the flexibility-related activities that are spread across the architecture.

This module protects the system using its second sub-module shown in Figure 3.4, from the bottlenecking effect by following a distributed architecture approach for different resilience task application. Another main factor addressed in this module is recovery. As the IoT infrastructure is wide and diverse, hence recovery activities must be automated and self-organized. This can be achieved here by utilizing the knowledge base of the infrastructure along with its state, and combining this information with some cognitive techniques, is capable of self-healing.

3.5.3.1.5 Other Management Functions

This module represents additional management functionalities that may be required by other layers. This module's components are not fixed. Based on the application, the management functions may vary. This type of tasks will be required by the IoT middleware. For example, Security Manager would be an essential component that could be needed by an agency. Its portion will tackle the dangers posed by the IoT to mitigate vulnerabilities. This section will attach sector basis services, not static ones.

3.5.3.1.6 Data Collection and Diffusion and Knowledge Database

The data collection and diffusion components are responsible for collecting data from sensors in the IoT system directly or via IoT gateway, i.e., fog/edge layer, and for disseminating this data to IoT services. On the other hand, the knowledge database preserves knowledge about the equipment and the physical topology of the system. This module will deal with the heterogeneity of the systems, and therefore it is important to use a standard to handle and identify these systems in a consistent manner. For this last part, an attempt could be made to use semantic web techniques, such as ontologies [21].

3.5.3.1.7 IoT Service Manager

This layer includes a structured way to access the IoT middleware features. It manages the overall services executed by the IoT middleware and stays with sync, with the response coming back from the IoT service layer. This part implements a specific application interface to allow for transparent communication of services to the cloud layer's upper portion, i.e., the IoT service layer of the unified architecture with the IoT middleware.

3.5.3.2 IoT Service Layer

The most significant and critical aspect of the architecture is the provision of end user services. Extracted data are stored in the cloud, and services are delivered to users in the form of microservices. Every complex system can be viewed as a cluster of microservices. These microservices are considered as an extension of SOA. Microservices enable the cloud to deploy loosely coupled services. This microservice concept applies to the Fog layer also. Because of their nature, microservices are becoming more and more useful for IoT architecture. As a result, IoT systems are evolving from "Things"-oriented ecosystems to broad and well-distributed micro-service-oriented ecosystems. Microservice has some very interesting built-in characteristics, e.g., self-containment, organized around business capabilities, monitoring and elimination of fault cascading, smart endpoints and dumb pipe for the handling of various service models, extensibility and container-based rapid deployment, lightweight in behavior, single activity orientation, decentralized governance, fragmented object prevention, decentralized data management, strengthened logging, powerful modularization, infrastructure automation, composability, and plug-and-play, to identify a few. These are very favorable qualities to choose microservices as a building block of the IoT service layer in the cloud. This IoT service layer, composed of such microservices, often referred to as the computing layer, consists of a computational function for delivering cloud-to-end service. After middleware controls interoperability by handling heterogeneity, the service request arrives in this layer. After storing all of the necessary data in the cloud, this layer provides microservices using the API gateway. After the computation is performed by the microservice, the response goes back to the IoT service manager, which is in the IoT middleware for trouble-free activity in the cloud layer.

We need cloud storage here to be more reliable and to provide both individual and aggregated services, as well as data collected from various applications. To order to deliver a benefit to the customer, the processing is conducted in the cloud using the following elements [3].

Database: Filtered data collected from all the subsystems are ingested into the cloud. The database is used for storing the data in a more organized way to ease the data fetching. This component is also used to map the data storage as well as the results of the analyzing components with the stored data.

API gateway: It is the responsibility of the IoT service layer API gateway in unified architecture (Figure 3.5) to provide an API management solution. This serves as a single point of entry into the network, which provides APIs that execute different functions and communicate with the end user. We need to build REST APIs using this part. Using these APIs, http requests are submitted to the cloud from end user point applications, and events are triggered. So the event controller in the cloud

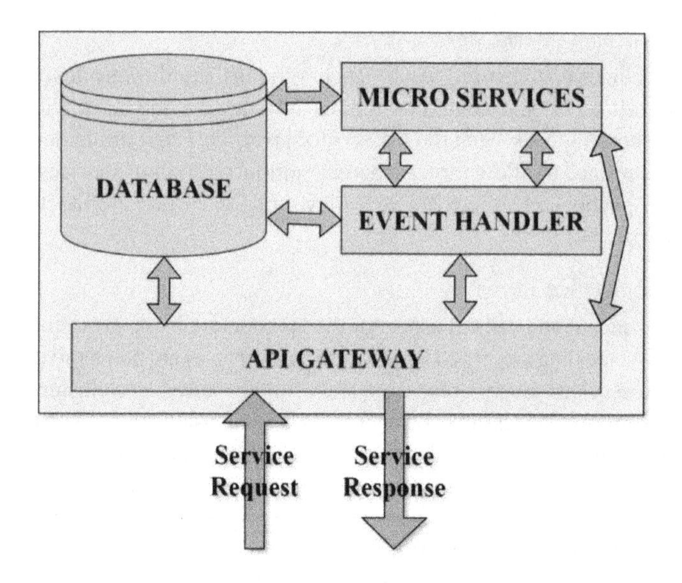

FIGURE 3.5 IoT service layer components for cloud layer.

initiates appropriate retrieval, allowing the functions to enter the application servers and data storage with the data needed to fulfil the demands. After the processing is over, http responses are given back from the event handler to the smartphone applications, through these APIs.

Event handler: The event handler is one of the unified architecture's main elements, and is essential for computing purposes. It includes individual functions for processing the data from an application's subsystems. It also contains single as well as fusion functions that use more than one subsystem's output data (or from more than one source) and analyzes it to obtain more statistical information. Once events are triggered by the above elements, the event handler runs certain functions that use microservices to obtain the results generated from stored data, and the result is then stored in the cloud data storage as needed. Depending on the triggered case, it can save the results for future use in the database, or use the API to provide the front-end user with a real-time response.

Microservices: Among the back-end services that are being provided, this component is responsible for different microservices, which include different models including machine learning, predictive analysis, deep learning, etc. When some tasks are required to be performed, this component gets triggered through the event handler and the results get stored in the database and used when required.

3.6 IoT-BASED REAL-WORLD APPLICATIONS AND UNIFIED FRAMEWORK

Now, to map real-world applications to the mentioned unified architecture, we have considered three different types of applications, i.e., IoT-based applications, smartphone-based applications, and a combination of both. However, among these three types of representative applications, the key difference in the orchestration of this architecture is in the existence of an IoT gateway (i.e., edge/fog layer). This IoT gateway (edge layer) can be used effectively for localized analytics and IoT-enabled interface for real-time control. However, in the case of either IoT and smartphone-based applications or only smartphone-specific applications, the local computation can be performed within smartphones as they are used as edge computing nodes. The reason for doing so is that smartphones now have a high computational capacity and relatively higher battery life. Besides, nowadays a smartphone has enough memory to collect its own sensed data or may collect nearby IoT device's data so that it can store data within itself for a short period before transferring the data through its internet connection to the cloud. For smartphones, too, power management can be performed efficiently [22]. For these applications, we have the data analytics section that usually takes place in the computation layer, which is in the cloud. This layer has specific functions for various types of IoT-based applications. Not all applications have the same components in the cloud layer for data analysis and computation. The cloud layer manages IoT-based data layer's heterogeneity; i.e., it manages the interoperability virtually in the IoT middleware between different IoT devices. This chapter already describes all of these layers and their functionalities before. Now we'll map these three real-world applications in this unified framework to understand the potential of its implementation.

Actually, most IoT-based devices are somehow connected to the smartphone. For different real-life applications, we consider applications with the smartphone as well as without a smartphone. Here, the presence of a smartphone in the architecture implies that the architecture utilizes the smartphone's functionality for either computation or data forwarding purposes. Smartphone-based application means that the end node in the system is a smartphone and its sensing capability is utilized in the IoT-based data layer. NoiseSense [23] is an example of that. For smartphone-based applications that include IoT-related applications, as well as where a smartphone is used for computational purposes as a part of the network, there is no need for a separate fog node in the architecture as described earlier, as computation can be done on the smartphone. An example of such an application is AirSense [24, 25]. Here, the smartphone is not a gateway but is required for data forwarding purposes. Even if the fog is present in such a structure, it will work as an internet gateway. In a full IoT-based application, e.g., smart building [26] application, a smartphone is not generally used as a computational node or an internet gateway for the IoT-based system, but rather used as an end node for device controlling purpose, i.e., the smartphone is only used as an end device to monitor and control IoT-based service at the user end.

Now we map these applications in the mentioned unified architecture for IoT. First, we take the smart building application where IoT is used to make building facilities and other household appliances IoT-enabled. In real-life applications, these objects/devices connect directly to smartphones using its Bluetooth if devices are within its range or otherwise via nearby internet gateway devices. In both cases, these things are connected with the internet. The prototype is built to control buildings' electrical appliances remotely along with plenty of other smart features that are also available in the building. The smart building system's realization follows a three-tier architecture (IoT slicing), as shown in Figure 3.3. In the data layer, we have all the IoT-based solutions that are mentioned in the smart building [26], which are forming the data layer. Then, the data is using an IoT gateway (Fog Router), which is doing the local processing for a quick response. From the gateway, data is forwarded to the cloud layer. In the cloud layer, IoT heterogeneity is handled in IoT middleware. Then the service request is further forwarded in the IoT service layer where microservices are modelled. This layer generates the response for any specific service requested by the user. This response comes back to the user end through IoT service manager to the user's smartphone.

Similarly, for other smartphone-based applications, AirSense or NoiseSense is mapped with the same generalized architecture. In AirSense, data are collected using IoT-based device AQMD, and then the data are forwarded to the smartphone. The smartphone adds more information, which it senses using the smartphone's system with the AQMD's data, and creates a valid data tuple and each device is uploaded the same for further analysis in the cloud. AirSense provides multiple services to users. Those microservices are requested by the AirSense users, which comes to the users as a service request response of the IoT service layer in the cloud. A similar thing is applicable for NoiseSense also. The main difference between these two applications is that NoiseSense is using the smartphone itself as the sensing as well as local computation device, whereas in AirSense, it is using the smartphone mainly for data forwarding purposes. Hence, in NoiseSense both data collection and pre-processing are happening in the mobile phone itself. After this, the collected data are pushed to the cloud. Thus, the smartphone is responsible for both data collection and data pushing in the cloud. Here, the functionally predominant layer is the data layer, which collects data using crowd-sensing. Local computation is done on the smartphone itself and it is also used as an IoT gateway. The rest of the data travel through the unified architecture is very similar, as mentioned in the AirSense application. Thus, all these three real-life applications mentioned are very much feasible to implement as well as to scale up with this architecture, though the types of the applications are different. Hence, we can say this architecture can handle data heterogeneity as well as different types of applications as well.

3.7 MAPPING OF BIG DATA FRAMEWORK FOR IoT TO UNIVERSAL FRAMEWORK FOR POTENTIAL APPLICATIONS

The rise of data generated by interconnected IoT systems has played a significant role, impacting all areas of technology and industry by increasing the opportunities in the Big Data environment for organizations and individuals. The rapid growth of IoT in all industries indicated that the data are increasing rapidly. Since IoT devices can generate almost all the data formats, hence tackling that diversity is difficult. The form of the data is also not regular; we usually get a mixture of semi-structured, unstructured as well as structured data in different types of IoT-based applications. Different IoT-based applications generate different types of data including audio, video, xml, JSON, csv, etc., and the database stores all these types of data. As the applications scale up, so does the data. Big Data typically requires huge storage, enormous processing power, and may result in high latency. Such problems require different processing and computation layers of Big Data within the IoT chain. A closer

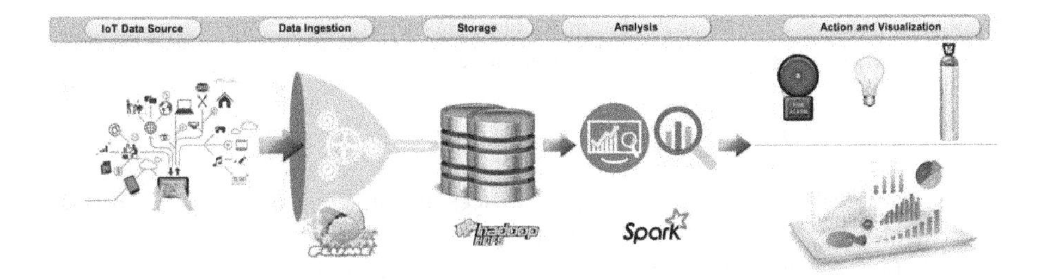

FIGURE 3.6 IoT-Big Data architecture.

inspection of IoT also revealed problems with scalability, latency, and bandwidth, as well as privacy, protection, availability, and durability control. The complexities in managing Big Data are important as the overall performance is directly proportional to the data processing service's properties. Analyzing or mining vast quantities of data generated to extract useful knowledge from both IoT applications and current IT systems requires strong Big Data analytics skills, which may be difficult for many end users in their implementation and interpretation. Our primary aim here is to map this service with our unified platform that can manage not only small-scale data but also large-scale data. Integrating IoT devices with external resources such as existing software and web services requires different middleware solutions to be created, as applications can vary significantly with industries. Here, we map the Big Data framework presented in [27] to this unified framework for IoT.

The architecture shown in Figure 3.6 is a generalized one that can handle IoT-based Big Data, and allows the generation, extraction, and ingestion of various types of IoT data in HDFS. It creates informative visualization and real-time analysis with Machine Learning and other state-of-art models that generate meaningful information and good decision making. The design's commitment is to make a novel architecture with features such as variety, flexibility, low costs, and quick decision making. Since using the Cloudera Distributed Hadoop (CDH) platform creates an advantage in storage costs, the versatility of integration with many technologies that allowed them to collect data and manage their information is low in cost. Because of Hadoop's usage, a decision is taken faster, which speeds up data analysis and processing of the knowledge. New data sources are taken care of faster in the system. Thus, these processes help businesses by providing immediate knowledge regarding the data and help them to make informed decisions. Now, to map this with our unified architecture, the layer of data generation here is the same as the layer of IoT, which generates heterogeneous data, i.e., Big Data. The Fog layer does its job accordingly based on the need. Then the collected data are ingested in the cloud. Heterogeneity of the data is managed by the IoT middleware in the cloud. IoT Services part does the data analysis part and proper action is taken based on the analysis outcome. Here, all the components are the same in the abstract level, but when you go into the details, you will find that only technology will be different for different applications based on their need. Here, [27] authors have used Flume, HDFS, Spark, etc., for the realization of this IoT-Big Data generic architecture, which will be different for other applications as well in the unified architecture. Hence, unified architecture maps and fits well in this IoT-Big Data generic model as well.

3.8 CHALLENGES

IoT has built a vast network of connected devices that are constantly interacting with millions of other "Things" linked to the internet. This progress of the IoT industry is overwhelming. It has plenty of positive impacts on the human population. It brings humans even closer to technology than before. Now, in terms of the industry also, it has a significant impact. It works as a catalyst for the long-awaited Industry 4.0 revolution. Nevertheless, the development of IoT has been hampered by the lack of regulation in the sector, which is still considered in its infancy [28]. The International Organization for Standardization (ISO) has carried out plenty of research with its scientists for over a decade to determine a common standard for IoT. They have also studied regarding financial benefits associated with the standardization process of new technologies and it is huge [29, 30]. Other regulatory bodies, e.g., ITU, IEC, and IEEE have also contributed a lot to establish a common standard [31]. But still, no unique standard is present for the IoT framework, thus the industry suffers from unregulated IoT sprawl. Hence, the presence of specific guidelines, regulation, and universal standards are necessary to create a healthy industry, which in turn can create a cost-effective useful secure product in the mass scale for the consumer.

It is essential, when discussing standards, to first understand the meaning and aim behind their implementation. A standard's primary objective is to systematize instruction implementation. Standards developed by studies, and endorsed by industry experts, are both authoritative and reliable. A standard is intended to formalize an acceptable way to do things, create a product, measure an item, or provide a service. And in this way, guidelines can enable companies to systematically and universally implement guidance across the board. Furthermore, in later stages of production, the company will evaluate its production standard with the original standard mentioned by the standardization authority. In different fields, there are plenty of global standards that are present already. Which came to the market after a comprehensive scientific study, which is a partnership between industry, academics and legislative bodies, effective standards are developed. When finalized, companies must then follow standards. Industry expert expertise plays a key role in developing successful standards. Putting a standard in the field not only improves the product's quality and service, but it also helps the expansion of scientific knowledge in that field which leads to a new direction for further research.

Within IoT systems, there is a complex and distributed collection of networks, modules, and applications that serve trillions of IoT artefacts. All these areas of IoT systems operate without any uniform coherent structure. The massively beneficial role that norms have played in advancing culture, technology, and the economy is immense. With the previously mentioned benefits, standards often have a secondary function to raise awareness of possible hazards that may occur as a result of utilizing incompatible manufacturing and production methods. Additionally, standards also provide insight into different areas of operation, production, and maintenance for new and emerging organizations. Otherwise, due to lack of resources or technological knowledge, this experience may not be possible for them. The rules laid down in standards include guidance on requirements and different component varieties, which can be integrated into products. Compliance with a particular standard can be an effective route for manufacturers to gain reputation and confidence with all vendors, distributors, and end users. Consumers and end users typically see manufacturers who meet with the standards as being more trustworthy. Compliance is seen by end users as an indicator that goods have been put through tried and tested production processes and testing. Besides, compliance with standards ensures that the final product is designed to good quality and will work according to the requirements given. In assurance, it gives the customer the trust that the product is safe, reliable, and will work as expected. Another additional advantage of quality compliance leading to certification improves organizations and product image. This reputation will help organizations to grow, succeed, and achieve their financial goals.

Therefore, the presence and adherence to guidelines helps to minimize carbon emissions, injuries, accidents, improved credibility of the company, and a positive economic impact. Now the emphasis should be changed from general standardization to the IoT industry instead. The system serves as a complex global network backbone, which is seamlessly linked to other knowledge networks. IoT's long-term sustainability depends heavily on successful and interoperable standards that will cover all aspects of IoT development and ongoing life cycle, from product design and production to repair, warranty and software upgrades, maintenance, decommissioning, and disposal. The introduction of standardization would also improve cooperation on security and other technology fronts, using a universal standardized framework for ecosystem application development. Likewise, IoT control in the manufacturing, delivery, and maintenance processes must ensure the protection and integrity of the device, as well as end user privacy and confidentiality. Therefore, standardization will offer much-needed control and sophistication to address security and reliability issues currently affecting IoT, which is urgently necessary. To summarize, standardization eliminates technological barriers to trade, enables interchangeability that leads to increased competitiveness, lower production costs, increased process performance, and also offers customers the opportunity to conduct comparative analysis and review product compliance with local regulations before making a rational purchasing decision.

The entire IoT industry faces massive challenges, which are growing in magnitude and evolves. IoT system vulnerabilities provide easy access to attackers, which lead to more malicious attacks, data theft, data destruction, hardware damage for the IoT devices, as well as the framework. Also, owned devices enable organized large-scale attacks on IT infrastructure, which can be experienced by consumers across geographical boundaries. Below are some of the big issues impacting IoT.

3.8.1 Security

The IoT industry remains unregulated, which has had broader consequences for security and privacy. A brand new attack vector has been created by the widespread adoption of unsecured IoT devices in almost every industry, including health and military, and by its increasing popularity among end users. The simplicity of accessing IoT systems, sniffing packet data, and modifying unsecured firmware is frightening.

3.8.2 Interoperability

Interoperability among IoT devices is severely constrained due to the profusion of components, data layers, languages, and supporting hardware and software involved in the development of an IoT device. In an ideal framework, the services mentioned above would easily work together to assist connectivity and knowledge exchange. Communication interoperability in the IoT arena, however, is especially challenging due to the wide range of available technologies, which poses a challenge for seamless interaction between multi-devices. Accordingly, standards which include interoperability are urgently required.

3.8.3 Governance

Effective governance within the IoT industry is an important consideration. Because of IoT's entrance into almost every sector, malicious backdoor passages left by insecure devices might not be instantly evident, but it opens up even larger-scale new paths for cyber attacking. Global regulation, such as rules, processes, protocols, audits, transparency, and continuity, is also currently non-existent in the IoT domain, due to the lack of general legislation in the IoT sector. These levels of business, regional, and international regulation can be extremely helpful in providing companies with enhanced system performance and reliability, as well as reducing the risk of potential errors.

3.9 SUMMARY

We are surrounded in the present day and age by IoT products. It has a very diverse field of application, which is smart and user-friendly. We do know plenty of applications, for example, it's not unusual today to see lights that can automatically turn on when sensing someone's presence, or heating, ventilation, and air conditioning (HVAC) systems that can toggle when the temperature drops or reaches the desired level. Likewise, we already have devices on the market such as smart kettles that can be triggered remotely via a smartphone, or location-aware thermostats that can switch on the heating when they notice that the resident of the home leaves the office for home. Equally, always-on voice-activated products with virtual assistants are also becoming increasingly popular. These apps provide accurate voice recognition, allowing the virtual assistant to support the user in day-to-day activities. Comparably, our cities are now becoming smart, for example, now IoT devices are being used to control parking, monitor traffic congestion, and track the movements of people, enabling the study to change the city crowd's habits. Smart sensors are used in commercial industries to monitor and maintain equipment, climate, including temperature within medical storage facilities or medical vehicles, and to track important or valuable commodities.

Now, to manage these whole sets of applications, a unified framework is required, which can bundle all the applications under one umbrella, such that dominant problems are tackled uniformly. Here, we have studied different research articles and have come up with a valid unified architecture which can incorporate all the applications. This architecture is also validated with different real-world applications. The problem that remains, even after the framework, is standardization, which is missing. Different international organizations are trying their level best to standardize it, but until now, we don't have the standard. Based on the framework described in this chapter, if standardization is implemented on different layers, then the issue of the framework uniformity can be shed light. Today's biggest issue with standardizing IoT framework is that the IoT needs a different approach than a normal system. These problems are currently being considered with the highest priority. Most of the standards either concentrate on general issues or consider their particular topics. However, it is anticipated that the probability of full compliance in the industry is likely to be small for some time, even after the regulations are enforced. It is suggested, to encourage compliance, that governments would need to introduce compliance incentives. Practically, compliance with standards will increase trust between consumers.

REFERENCES

1. Statista Research Department (Feb 19, 2020), Number of internet of things (IoT) connected devices worldwide in 2018, 2025 and 2030, https://www.statista.com/statistics/802690/worldwide-connected-devices-by-access-technology/
2. F. Khodadadi, A.V. Dastjerdi, and R. Buyya, Chapter 1—Internet of Things: An Overview, pp. 3–27, 2016. ISBN 9780128053959, https://doi.org/10.1016/B978-0-12-805395-9.00001-0.
3. A.I. Middya, S. Roy, J. Dutta, and R. Das, JUSense: A Unified Framework for Participatory Based Urban Sensing System. In: *Mobile Networks and Applications*, Springer, Berlin/Heidelberg, 2020.
4. J. Huang, Q. Duan, C. Xing and H. Wang, "Topology Control for Building a Large-Scale and Energy-Efficient Internet of Things," *IEEE Wireless Communications*, vol. 24, no. 1, pp. 67–73, February 2017.
5. G. P. Jesi, E. Benetti and G. Mazzini, "Building an IoT Public Network Infrastructure," *2019 International Conference on Software, Telecommunications and Computer Networks (SoftCOM)*, Split, Croatia, 2019, pp. 1–5.
6. L. Zhao, W. Sun, Y. Shi, and J. Liu, "Optimal Placement of Cloudlets for Access Delay Minimization in SDN-Based Internet of Things Networks," *EEE Internet of Things Journal*, vol. 5, no. 2, pp. 1334–1344, April 2018.
7. S. Bera, S. Misra and A. V. Vasilakos, "Software-Defined Networking for Internet of Things: A Survey," *IEEE Internet of Things Journal*, vol. 4, no. 6, pp. 1994–2008, December 2017.
8. M. Barcelo, A. Correa, J. Llorca, A. M. Tulino, J. L. Vicario, and A. Morell, "IoT-Cloud Service Optimization in Next Generation Smart Environments," *IEEE Journal on Selected Areas in Communications*, vol. 34, no. 12, pp. 4077–4090, December 2016.
9. L. Toka, B. Lajtha, É. Hosszu, B. Formanek, D. Géhberger, and J. Tapolcai, "A resource-aware and time-critical IoT framework," *IEEE INFOCOM 2017—IEEE Conference on Computer Communications*, Atlanta, GA, 2017, pp. 1–9.

10. J. Dutta, S. Roy, and C. Chowdhury, "Unified Framework for IoT and Smartphone-Based Different Smart City Related Applications," *Microsystem Technology*, vol. 25, pp. 83–96, 2019. https://doi.org/10.1007/s00542-018-3936-9
11. K. Yelmarthi, A. Abdelgawad, and A. Khattab, "An Architectural Framework for Low-Power IoT Applications," *2016 28th International Conference on Microelectronics (ICM)*, Giza, 2016, pp. 373–376.
12. I. Ganchev, Z. Ji, and M. O'Droma, "A Generic IoT Architecture for Smart Cities," *25th IET Irish Signals & Systems Conference 2014 and 2014 China-Ireland International Conference on Information and Communications Technologies (ISSC 2014/CIICT 2014)*, Limerick, Ireland, 2014, pp. 196–199.
13. S. Cheruvu, A. Kumar, N. Smith, and D. M. Wheeler, IoT Frameworks and Complexity. In: *Demystifying Internet of Things Security*. Apress, Berkeley, CA, 2020. DOI: https://doi.org/10.1007/978-1-4842-2896-8_2
14. S. K. Datta and C. Bonnet, "Next-Generation, Data Centric and End-to-End IoT Architecture Based on Microservices," *2018 IEEE International Conference on Consumer Electronics - Asia (ICCE-Asia)*, Jeju, 2018, pp. 206–212.
15. David Perez Abreu, Karima Velasquez, Marilia Curado, and Edmundo Monteiro, "A Resilient Internet of Things Architecture for Smart Cities," *Annals of Telecommunications*, vol. 72, pp. 19–30, 2016. https://doi.org/10.1007/s12243-016-0530-y.
16. LinkSmart (2020), LinkSmart middleware platform, https://linksmart.eu/. Last visited: 2020-04-09.
17. OPENIoT (2020), OPENIoT—open-source cloud solution for the Internet of Things, http://openiot.eu. Last visited: 2020-04-09.
18. Rost, Matthias Johannes, "Virtual network embeddings: theoretical foundations and provably good algorithms." Dissertation, 2019.
19. D. Zeng, L. Gu, S. Pan, and S. Guo, Software Defined Networking II: NFV. In: *Software Defined Systems: Springer Briefs in Computer Science*. Springer, Cham, 2020.
20. T. Adesina and O. Osasona, "A Novel Cognitive IoT Gateway Framework: Towards a Holistic Approach to IoT Interoperability," *2019 IEEE 5th World Forum on Internet of Things (WF-IoT)*, Limerick, Ireland, 2019, pp. 53–58.
21. Gergely Marcell Honti and Janos Abonyi, "A Review of Semantic Sensor Technologies in Internet of Things Architectures," *Hindawi Complexity*, 2019. https://doi.org/10.1155/2019/6473160
22. J. Dutta, P. Pramanick, and S. Roy, Energy-Efficient GPS Usage in Location-Based Applications. In: Satapathy S., Tavares J., Bhateja V., Mohanty J. (eds) *Information and Decision Sciences. Advances in Intelligent Systems and Computing*, vol. 701. Springer, Singapore, 2018. https://doi.org/10.1007/978-981-10-7563-6_36
23. J. Dutta, P. Pramanick, and S. Roy, "NoiseSense: Crowdsourced context aware sensing for real time noise pollution monitoring of the city," *2017 IEEE International Conference on Advanced Networks and Telecommunications Systems (ANTS)*, Bhubaneswar, India, 2017, pp. 1–6. DOI: 10.1109/ANTS.2017.8384103
24. J. Dutta, F. Gazi, S. Roy, and C. Chowdhury, "AirSense: Opportunistic Crowd-Sensing Based Air Quality Monitoring System for Smart City," *2016 IEEE SENSORS*, Orlando, FL, 2016, pp. 1–3. DOI: 10.1109/ICSENS.2016.7808730
25. Joy Dutta, Chandreyee Chowdhury, Sarbani Roy, Asif Iqbal Middya, and Firoj Gazi. "Towards Smart City: Sensing Air Quality in City based on Opportunistic Crowd-sensing." *Proceedings of the 18th International Conference on Distributed Computing and Networking (ICDCN'17)*, ACM, New York, NY, 2017, Article 42. DOI: https://doi.org/10.1145/3007748.3018286
26. J. Dutta and S. Roy, "IoT-fog-cloud based Architecture for Smart City: Prototype of a Smart Building," *2017 7th International Conference on Cloud Computing,*

Data Science & Engineering, Noida, India, 2017, pp. 237–242. DOI: 10.1109/CONFLUENCE.2017.7943156

27. Fabián Constante, Jorge-Luis Pérez-Medina, and Paulo Guerra-Terán, "A Proposed Architecture for IoT Big Data Analysis in Smart Supply Chain Fields." *The International Conference on Advances in Emerging Trends and Technologies (ICAETT 2019)*, Springer, Cham, October 2019.
28. Jibran Saleem, et al. "IoTstandardisation: Challenges, Perspectives and Solution," *Proceedings of the 2nd International Conference on Future Networks and Distributed Systems*, 2018.
29. International Organisation for Standardisation (ISO) Good Standardization Practices (2019). https://www.iso.org/files/live/sites/isoorg/files/store/en/PUB100440.pdf
30. ISO/IEC 30141:2018(en) Internet of Things (IoT) Reference Architecture (2018). https://www.iso.org/obp/ui/#iso:std:iso-iec:30141:ed-1:v1:en
31. N. Miloslavskaya, A. Nikiforov, K. Plaksiy, and A. Tolstoy, Standardization Issues for the Internet of Things. In: Rocha Á., Adeli H., Reis L., Costanzo S. (eds) *New Knowledge in Information Systems and Technologies (WorldCIST'19). Advances in Intelligent Systems and Computing*, vol. 931. Springer, Cham, 2019.

4 IoT Middleware Technology

Review and Challenges

Arindam Giri
Haldia Institute of Technology, West Bengal, India

Subrata Dutta and Kailash Chandra Mishra
National Institute of Technology, Jamshedpur, Jharkhand, India

Sarmistha Neogy
Jadavpur University, West Bengal, India

CONTENTS

4.1 Introduction 72
4.2 Use Case: IoT Middleware 73
4.2.1 Things Layer 74
4.2.2 Local Gateway 74
4.2.3 Internet 74
4.2.4 The Cloud 74
4.3 IoT Architecture 75
4.4 Enabling Technologies 76
4.4.1 Identification, Sensing, and Communication Technologies 76
4.4.2 Middleware Technology 76
4.4.2.1 Protocols Interface 77
4.4.2.2 Device Abstraction 77
4.4.2.3 Central Control, Context Detection, and Management 77
4.4.2.4 Application Abstraction 77
4.5 Review of Middleware 79
4.5.1 AURA 79
4.5.1.1 Task Manager (Prism) 79
4.5.1.2 Service Supplier (SS) 79
4.5.1.3 Context Observer (CO) 80
4.5.1.4 Environment Manager (EM) 80
4.5.2 HYDRA 80
4.6 Applications and Social Impact of IoT 80
4.6.1 Smart Environment (Homes, Buildings, Office, Plant) 80
4.6.2 Healthcare 80

4.6.3 Smart Cities 81
4.6.4 Transportation 82
4.6.5 Social Connectivity 82
4.6.6 Agriculture 82
4.6.7 Pharmaceutical Industry 82
4.7 Challenges and Research Directions 83
4.7.1 Massive Scaling 83
4.7.2 Security and Privacy 83
4.7.3 Interoperability 84
4.7.4 Big Data and Its Management 84
4.8 Conclusion 84
References 85

4.1 INTRODUCTION

"Internet of Things" (IoT) is a worldwide network of interconnected devices relying upon the infrastructure of Information and Communication Technologies (ICT). It connects objects like mobiles and refrigerators. There is a big market opportunity for device manufacturers, service providers, and application developers. By 2020, billions of smart objects are expected to join in IoT. In [1], the market shares of IoT by 2025 are depicted. Among the other sectors, healthcare and related sectors possess majority share, while the rest of the share is occupied by electricity, agriculture, and security. Devices like refrigerators or washing machines will act like smart devices with computing and communication capabilities. They can be controlled from our offices. Not only that, the interconnected devices will communicate with each other over the internet to yield a pervasive computing environment. However, there are many challenges to confront because of the robustness and heterogeneous nature of IoT devices. India and similar countries can be benefited with an application in agriculture using IoT. Using the applications, farmers can monitor the different field parameters, like moisture and temperature. It does not need physical supervision of the crop field. Even farmers can get expert advice about harvesting or irrigation. Such an application not only increases quality production, but also can save natural resources, like water. Healthcare services can be offered to far-apart patients with IoT too. Devices attached to a patient send data like sugar level to a doctor. After getting various data pertaining to a patient, the doctor can advise medicines via patient's smart phone. This application relives the patient's energy and keeps hospitals free of crowds. Real-time tracking of goods is also possible nowadays. Goods can be attached with sensors which are aware of location. Through IoT, we can track the location of the goods and can estimate delivery thereafter. Smart meters can send reading or even usage pattern of a city to an electric grid. The grid now is able to distribute electricity efficiently over many cities. Animals like tigers are attached with radio-frequency identification (RFID). So, through IoT their movement can be traced. Medicines can be attached with smart labels so as to read information regarding date of expiry, doses, price, etc. In this way, counterfeiting of medicines can be prevented. Another contribution of IoT is ambient assisted living (AAL). Elderly people living alone in houses can be monitored remotely using IoT. Their near ones can be notified on emergency after analyzing their condition. Some of the home appliances, like air conditioning, can be automatically turned on or off remotely over the internet. IoT has been successfully used in smart cities. Traffic lights, parking space, and water distribution, etc., in a smart city are controlled by the use of IoT. The popularity of IoT is made possible with the advancement in enabling technologies of IoT, including identifying, communication, sensing, and the middleware. A software layer called middleware is positioned between the IoT technical and application layers in order to balance the technical gaps between different manufacturers [2–5]. This chapter presents a comprehensive overview of IoT, its architecture, enabling technologies [6], and the state-of-the-art of middleware technology, along with research challenges for IoT.

The remainder of this chapter is structured as follows: Section 4.2 gives the use case of IoT middleware, while the architecture of IoT is mentioned in Section 4.3. Enabling technologies of IoT are presented in Section 4.4. A review of IoT middleware is included in Section 4.5. In Section 4.6, promising applications of IoT are described. Current research challenges are pointed out in Section 4.7. Finally, a conclusion is made in Section 4.8.

4.2 USE CASE: IoT MIDDLEWARE

In this section, we try to illustrate the motivation behind IoT middleware. The aim is achieved by developing an application framework for smart agriculture, such as the AgriTech [7] as given in Figure 4.1. Such a type of application has immense importance in third-world countries, where common people depend on agriculture. In traditional agriculture, the farmers need to observe the field closely. It requires the farmer's direct intervention. However, today people are busy in other professions as well. So, they don't have time to visit a crop field physically. As smart phones become popular to everyone, irrespective of professions, for field monitoring physically reaching to the field is not required. Instead, field data can be obtained from the cloud using AgriTech.

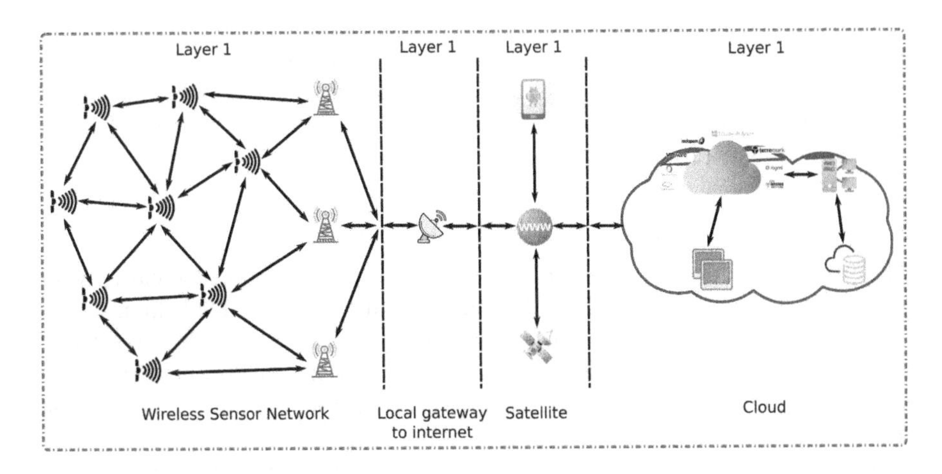

FIGURE 4.1 Use case: AgriTech framework [7].

As proposed by authors in [7], the framework for such an application has four layers:

4.2.1 Things Layer

Different devices are deployed in this layer with an aim to collect field data. For example, moisture sensors may be deployed to read the soil moisture. In a large field, many moisture sensors are used. They sense moisture and send to local base stations via networking among sensors. Wireless sensor network plays an important role in this context. It is a network of sensors to gather sensed data at base station either by directly or by multi-hop way.

Knowing the location of sensed data, it is essential to take decisions based on those data. The simple way to achieve location of every sensor is to attach Global Positioning System (GPS) to every sensor. But, this is not feasible as it increases application cost. Only a few sensors are attached with GPS, called anchors. All other sensors calculate locations using the locations of anchors. This process is known as localization. Here, in this use case, the authors consider to utilize a localization algorithm for the purpose of knowing humidity data. After knowing humidity with location information, the farmer can decide to start controlled irrigation in the specific field area. Not only does it restrict wastage of water, but it also saves time to manually check the dried-up field area.

4.2.2 Local Gateway

Data can be collected from sensors either by directly or multi-hop way. Local base stations collect the data and forward to a local gateway. Its job is to aggregate the data to restrict redundancy of data. Once the aggregation is over, it uploads the same to the cloud. As the smart phones have limited storage capacity, field-related data can't be stored in such devices. We can use the cloud for storage requirement.

4.2.3 Internet

Without the internet, farmers can't access data saved in the cloud. Whenever they need to know different field data, the cloud is used and the data is downloaded in the smart phones of farmers. They can also get weather data from the internet. Weather data helps in taking irrigation or harvesting crops.

4.2.4 The Cloud

The cloud provides a storage facility in the framework. Farmers need to pay only for knowing field data from the cloud. If needed, farmers can take the help of advice from agricultural experts. They can provide irrigation or harvesting decisions based on field data. For example, analyzing data and considering weather conditions, the experts can suggest postponing the irrigation. This decision not only saves crops but also prevents wastage of water.

In the framework, there are many heterogeneous devices. In order to manage the technological differences, the middleware layer is to be introduced

4.3 IoT ARCHITECTURE

The architecture of IoT must support heterogeneous devices. Although several architectures are proposed, no one of them is chosen as a common architecture for wide acceptance. The existing TCP/IP protocol suite is not suitable for IoT because of scalability. Presently, new architecture needs to be developed to handle IoT network, addressing many challenges. As of now, many architectures have been proposed for IoT, as shown in Figure 4.2. The architectures may be classified as: application-specific and general purpose [8–11]. The standard architectures for a multifaceted concept are appropriate. A multi-layered architectural model may have 3, 4, 5, or 6 layers [5]. Among them, the Service Oriented Architecture (SOA)-based model is most popular. It is supported by middleware technology. A summary of different architectures is given in Table 4.1.

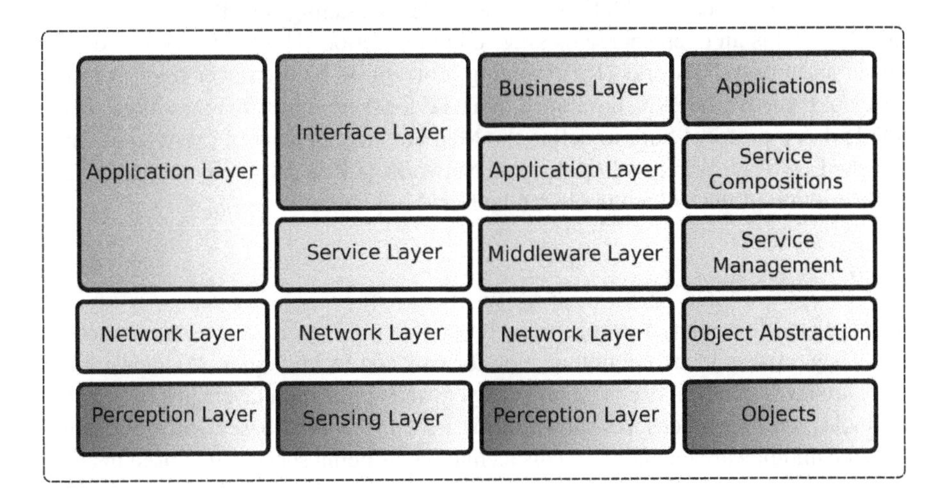

FIGURE 4.2 Layered Architectures: First column: three-layer based, Second column: four-layer based, Third column: five-layer based, Fourth column: the SOA-based architecture [5].

TABLE 4.1
Summary of Layered Architecture

Model	References	Security	Scalability	Reliability	QoS
Three-layered	[8]	-	-	-	-
Four-layered	[9, 10, 13]	✓	-	-	✓
Five-layered	[5]	✓	-	-	-
SOA	[14, 15]	✓	-	✓	✓

4.4 ENABLING TECHNOLOGIES

Technological changes in ICT make IoT realizable to us. This section includes the enabling technologies, which makes IoT feasible.

4.4.1 Identification, Sensing, and Communication Technologies

IPv4 and IPv6 are being used to address IoT devices. 6LoWPAN [12] is developed for low-power wireless communication over IPv6. The RFID [8] is applied to identify the objects attached with radios.

For communication, IEEE 80215.4, Bluetooth, IoT, and WiFi are commonly used. WiFi supports communication in short distance [11]. Near Field Communication (NFC) provides short- range communication of 10 cm and a data rate of 424 [16].

In addition to standards in Table 4.2, organizations like EPCGlobal provide efforts towards developing protocols. The Constrained Application Protocol (CoAP) [17–20] is developed by IETF. The Extensible Messaging and Presence Protocol (XMPP) [21] is an open source message communication. The IETF developed Low Power and Lossy Networks (RPL) routing protocol, to be used over IPv6. Another protocol is the 6LoWPAN, developed by IETF for communication over IPv6. IEEE 802.15.4 is a protocol built to define the Medium Access Control (MAC) sub-layer and the Low-Rate Wireless Private Area Networks (LR-WPAN) physical layer [22]. A description of the IoT protocols can be found in [7].

4.4.2 Middleware Technology

It's predicted that many heterogeneous objects would interact and communicate in IoT. They possess many communication formats and technologies. Middleware is a software layer added between technical layers and application layers just to meet the technological differences. As shown in Figure 4.1, the SOA-based IoT architecture uses the middleware technology. This technology is being successfully used in cloud

TABLE 4.2
IoT Communication Technologies [24]

Communication Protocol	Transmission Range	Transmission Rate	Spectrum
RFID	50 cm/50 cm/3 m/ 1.5 m	424 kbps	135 KHz/13.56 MHz/ 960 MHz/2.4 GHz
Bluetooth	10 m	1 Mbps	2.4 GHz
WiFi	100 m	50–320 Mbps	2.4/5.8 GHz
NFC	10 cm	100 kbps to 10 Mbps	2.45 GHz
ZigBee	10 m	256 kbps/20 kbps	2.4 GHz/900 MHz
Wi-Max	50 km	70 Mbps	2–11 MHz
UMTS/CDMA/EDGE	–	2 Mbps	896 MHz
IEEE 802.15.4	10 m	20/24/250 kbps	868/915/2400 MHz

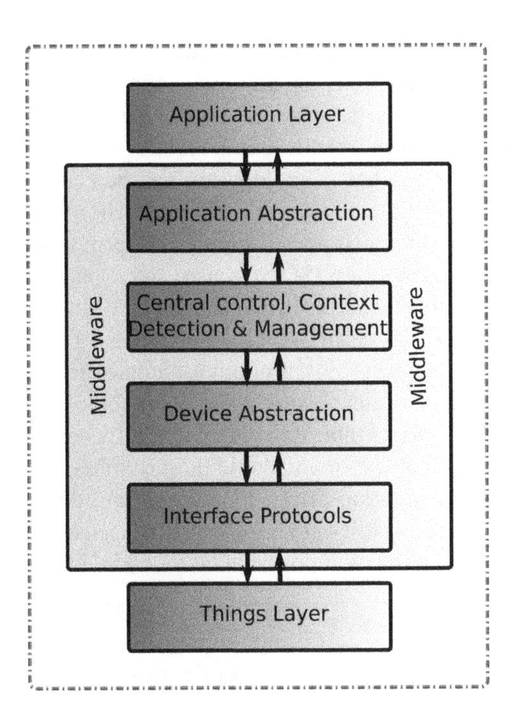

FIGURE 4.3 Functional components of IoT middleware.

computing, wireless sensor networks, and vehicular network [8, 23]. This section addresses the state-of-the-art of middleware technology.

The following blocks, as shown in Figure 4.3, are functionally an IoT middleware.

4.4.2.1 Protocols Interface

Specifies the protocol between networks to share information. This block handles connectivity issues in the TCP/IP protocol stack in physical layer to upper layers.

4.4.2.2 Device Abstraction

Describes an abstract system model for facilitating interaction between heterogeneous devices. APIs are responsible for managing syntactic or semantic of information. Syntactic means the format of the information, while the meaning of the exchanged information is reviewed by semantic interoperation.

4.4.2.3 Central Control, Context Detection, and Management

Manage context in IoT. Here, decision making is made involving contexts [25]. Currently, a few middleware support such computing in IoT. Research to contribute in the area of context-aware computing is being supported by the European Union during the period 2015–2020 [26].

4.4.2.4 Application Abstraction

Supports an interface to communicate among heterogeneous devices. Query language can help in defining the interface.

The middleware for IoT can be classified as functional, non-functional, and architectural [27]. Functional requirements are associated with managing resource and data, and non-functional requirements tackle QoS like security. Issues like interoperability and context-awareness are dealt with architectural requirements.

Again, the design approaches of middleware can be classified into many types with respect to design. Table 4.3 includes such types. Event-based middleware acts when an event occurs. Service-oriented middleware provides services to applications. Service-oriented computing (SOC) is associated on services. Virtualization of infrastructure is possible through Virtual Machine (VM)-based middleware [28]. Agent-based middleware dispatches different modules of applications to mobile agents. Fault tolerance is ensured by migrating modules to other agents. HYDRA [19], UBIWARE promotes context-aware computing. Tuple-space middleware, such as LIME [29], TS-Mid [30] use tuple data. Here, an SQL query may be used to know the application data. In database-oriented middleware solutions, the application uses SQL-like statements. Application-driven middleware are designed to support only specific applications. Such middleware solutions do not necessarily meet all requirements. The current middleware solutions can be found in Table 4.4.

TABLE 4.3
Middleware Design Approach Classification [24]

Type of Middleware	Examples
Event-based	Hermes [20], RUNES [31], EMMA [32], GREEN [33]
Service-oriented	Hydra [19], TinySOA [22], SENSEI [34]
VM approach	MagnetOS [35], Sensorware [28], TinyVM [36]
Agent-based	Smart messages [37], Agilla [38], UbiROAD [39]
Tuple-space approach	LIME [40], TS-Mid [41], A3-TAG [23]
Database approach	COUGAR [42], Sensation [43], TinyDB [44], KSpot+ [45]
Application-specific	MiLAN [46], MidFusion [47], AutoSec [48]

TABLE 4.4
Comparison of Solutions for Middleware [15]

	Features of Middleware				
IoT Middleware	Interoperability	Device Management	Portability	Context Awareness	Security and Privacy
HYDRA	✓	✓	✓	✓	✓
ASPIRE	-	✓	✓	-	-
UBIWARE	-	✓	✓	✓	-
UBIROAD	✓	✓	✓	✓	✓
GSN	-	✓	✓	-	✓
SIRENA	✓	✓	✓	-	✓
SOCRADES	✓	✓	✓	-	✓
WHEREX	✓	✓	✓	-	-

Among the middleware architectures [14], SOA-based IoT architecture attracts attention recently. However, in the current scenario, more research contribution is sought in this domain. A few solutions are reliable as they provide security and privacy. The Interoperability issue is not addressed by ASPIRE, UBIWARE, GSN. Context awareness is not found in most of the solutions. HYDRA and UBIROAD supports all features. The potential research should explore all issues of IoT.

4.5 REVIEW OF MIDDLEWARE

In this section we give insights into two middleware solutions, namely, AURA and HYDRA.

4.5.1 AURA

AURA collects data from devices and manipulates them. It provides few APIs to support data service to end users. It manages device heterogeneity using such APIs. With its middleware, users with heterogeneous devices can communicate with each other. The functional diagram is given in Figure 4.4.

The AURA middleware is composed of four blocks.

4.5.1.1 Task Manager (Prism)

Task Manager (Prism) allows users to use the application in a platform-independent manner. It manages the necessary changes so as to adopt with the new situation arising out of the operating system or other changes due to configurations. Users don't feel any changes due to location changes also. It appears such as there are no changes happening. AURA provides Microsoft Word when in Windows, while it offers Emacs in Unix system for editing texts.

4.5.1.2 Service Supplier (SS)

Service Supplier (SS) sends the user's request to specific software. For example, it sends request to Emacs to serve the request of the user for editing tests. It does the

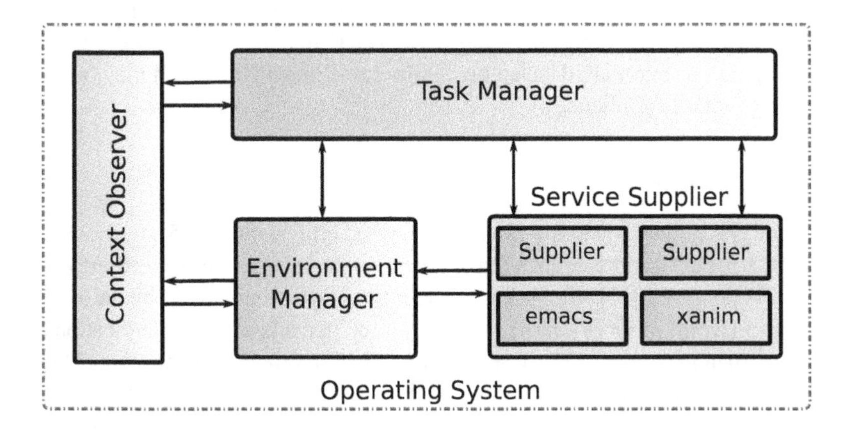

FIGURE 4.4 Functional blocks of AURA.

necessary setting to serve the user's requests. So, it acts as an interface to software services.

4.5.1.3 Context Observer (CO)

Context Observer (CO) collects information of users, like location and preference. Based on the context, it interacts with task manager or environment manager. Such operations are done automatically without the user's notice. The user is unaware of the operation or contexts.

4.5.1.4 Environment Manager (EM)

Environment Manager (EM) acts as a mediator among task manager, service supplier, and context observer. Depending on the service requirement, it interacts with a specific service supplier. It has the mapping from services to supplier.

4.5.2 HYDRA

One of the popular middleware solutions is HYDRA. It supports simpler configuration setting than others. This middleware is based on service-oriented architecture. So, mostly it shares the architecture of SOA-based architecture. The middleware of HYDRA is developed with intelligent software. It may be used in agriculture and healthcare.

The framework of HYDRA is given in Figure 4.5.

4.6 APPLICATIONS AND SOCIAL IMPACT OF IoT

IoT has many promising applications nowadays. Our lives can be changed with such applications. Here, we mention a few applications.

4.6.1 Smart Environment (Homes, Buildings, Office, Plant)

Every electronics devices are now having sensors. They can read many parameters in the device. For instance, water filters can send a message once it requires maintenance. Many of the daily applications can be automated with IoT. It can help an elderly man living alone by switching on or off if needed [49–51]. Industrial processes can also be automated based on parameters. Smart Home [52] may be used to control many home appliances.

4.6.2 Healthcare

Healthcare is a domain where IoT can make our life comfortable [53–55]. Different body factors of a patient can be collected and sent to a doctor for treatment. Staff patterns can also be observed [56]. Gathering inside a hospital is maintained by an IoT-based application. Health services can be provided to the patients at a remote location using community health services [57]. These patients need not to visit the hospital. They can be advised with message or mail by doctors. But this facility must preserve the patients' personal information. Another application is AAL for supporting elderly people [58, 59]. In [30], measures are included to protect patients' data. The eHealth [60] provides

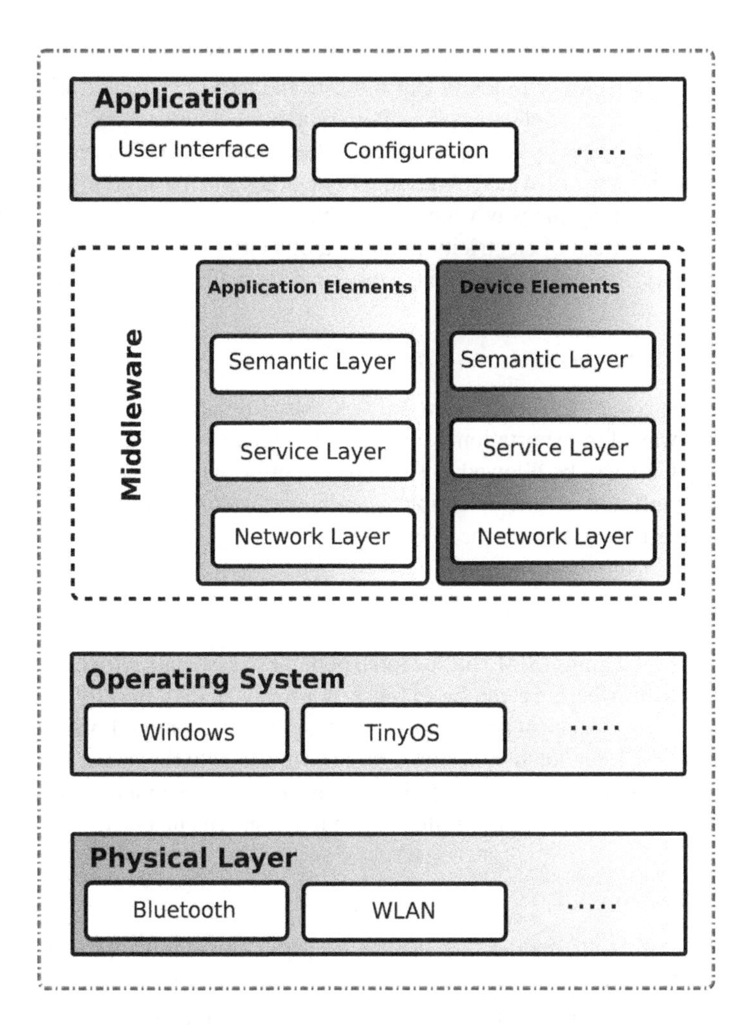

FIGURE 4.5 Framework of HYDRA.

advisory services to patients. Many other related services can be found in [61], and [62]. But we should be ready to face the issues in this area [63]. Devices are being developed by different vendors to support healthcare services [63–66]. As the healthcare information is very personal to patients, these informations need to be secured first. Patients may be unwilling unless their data is secured. Related issues in this regard may be found in [60]. Our country needs to develop proper monitoring of data security. Data centers must be able to access and digitize large-scale health data. As the application will be run on mobile devices, lightweight security measures should be developed.

4.6.3 Smart Cities

In order to enhance the life style of citizens in a city, this concept is developed using IoT, Resources in a city like water can be efficiently maintained by a smart city

[66–70]. It has been already applied in many cities over the world. It can help control traffic in a city. A traveler in a city can move to the parking space available at the moment using such an application [71]. Pollution can be better monitored by reading different parameters by sensors in a city [72]. Resources like water can be saved by controlling wastage of water. Water quality can also be ensured after deploying different sensors along the way water flows. Smart meters can be used in homes to know the usage pattern of electricity in an area of a city [73]. In spite of all benefits of the smart city, the data transfer with a city must be made secured from intruders.

4.6.4 Transportation

Transportation is another sector where IoT is successfully applied. It can monitor the goods on the way of transportation. Vehicles can be made so as to assist drivers while driving [74]. Cars can be allowed to share traffic information. Based on traffic, the driver can be advised to take diversion. The delivery status of goods is determined by tracking their location over the internet.

4.6.5 Social Connectivity

Social Internet of Things (SIoT) has become popular to associate with each other [2]. This is a platform where we can be in touch of near ones [53], in SIoT, we post our information so as to share with friends [54]. It is not only confined within humans, but in things also. Location of a lost thing can be informed to the owner by the developing application. Vendors are collecting feedback from customers to update their devices [55]. Providing security in SIoT is a major concern currently.

4.6.6 Agriculture

One of the main fields for the introduction of IoT is agriculture. In countries like Bangladesh, India, or Brazil, where agriculture is the main trade, IoT can be used and human effort can be optimally exploited. In [7] the authors suggested an agricultural automation system called AgriTech. Sensors such as moisture, nutrients, and data obtainment can save time from farmers in the agriculture sector. The excessive human effort in these countries can be used for industrialization. The potential impact of IoT in agriculture could be exploited by reducing waste of resources, such as water. In fields for sensing humidity, temperature, soil nutrients, and specific sensors are deployed. The amount of water, fertilizers, and insecticides is calculated on the basis of information from these sensors. This technique reduces agricultural costs.

However, the implementation of IoT is challenging in agriculture due to initial installation costs. Sensors are physically attackable in the crop field. Again, carefree sensor deployment generates unwanted field information that is not a farmer's part.

4.6.7 Pharmaceutical Industry

Health and protection are of the utmost importance for pharmaceutical products. In this view, intelligent drug labels are attached to their condition during transportation and stocking. With the understanding of medications, safety, and dosages, patients are specifically appreciated through clever labeling on products. Falsehood [75] can be stopped by IoT in this area. The cabinet of smart medicine helps patients to monitor medication delivery promptly. The IoT-based intelligent information system is presented to authors in [57], which helps medical practitioners recognize patient side-effects of different medicines.

A lot of data transactions will occur in any IoT system. The application will receive certain decisions and control signals or suggestions. A strong and effective authentication protocol is needed to ensure the protection of these data, suggestions, and control signals.

4.7 CHALLENGES AND RESEARCH DIRECTIONS

Although various projects have been done on IoT as shown in Table 4.5, still many challenges exist. Potential research issues are pointed out in this section.

4.7.1 Massive Scaling

IoT is densely populated with low-cost smart devices. Issues like identification, addressing, authentication, etc., are throwing challenges. Managing Big Data becomes a headache. Whether IPv6 is enough to address all devices? Can 6LoWPAN dominate in future internet? Moreover, a unified architecture is needed, supporting interoperability issue among heterogeneous devices.

4.7.2 Security and Privacy

Owing to the inherent openness of IoT, the risk of attack such as application attacks is present. Providing security is a concern in IoT [75]. Attack detection and resolving such attacks must be present in this situation. Lightweight security techniques may be applied [76].

TABLE 4.5
Comparison of Middleware Solutions [15]

	Features of Middleware				
IoT Middleware	Interoperability	Device Management	Portability	Context Awareness	Security and Privacy
HYDRA	✓	✓	✓	✓	✓
ASPIRE	-	✓	✓	-	-
UBIWARE	-	✓	✓	✓	-
UBIROAD	✓	✓	✓	✓	✓
GSN	-	✓	✓	-	✓
SIRENA	✓	✓	✓	-	✓
SOCRADES	✓	✓	✓	-	✓
WHEREX	✓	✓	✓	-	-

TABLE 4.6
IoT-Related Projects

Project	Website
Alljoyn [59]	https://allseenalliance.org
HYDRA [19]	www.hydramiddleware.edu
e-Japan strategy [77]	http://japan.kantei.go.jp/it/network/0122full_e.html
iCORE [78]	www.iot-icore.edu
IoT6 [79]	http://www.iot6.eu
IoTivity [80]	https://www.iotivity.org
IoT-A	www.iot-a.edu
SENSEI [34]	http://www.ict-sensei.org/index.php
RUNES [81]	www.ist-runes.org

In addition, the privacy of personal data in IoT must be preserved. For example, healthcare data of patients must not be shared with unauthorized users. There is a chance of violating the privacy of data in Big Data analysis in IoT. Social networking must preserve the personal information of users. The security measures in IoT are available in See Table 4.6 for more information.

4.7.3 Interoperability

Interoperability is the most wanted requirement among others for IoT. As the devices possess heterogeneous technical specifications made by different manufacturers, interoperability issues must be addressed for successful communication. APIs may help in designing interoperation among devices. APIs used in middleware can define syntactic and semantic interoperation.

4.7.4 Big Data and its Management

Large amounts of data can be communicated and stored in IoT, generating Big Data [82]. Big Data needs to be filtered and analyzed to interpret data. Data mining may be used to infer knowledge from Big Data. Most existing data mining algorithms are not suitable for Big Data due to their inherent centralized nature. The state-of-the-art of data mining techniques is found in [83]. As outliers deviate from other observations in data, detection and removal of outliers is a concern in Big Data [84], which are supported by [85] and [86].

4.8 CONCLUSION

IoT has become an inevitable part of our life with growing invasion of smart devices. Security and preserving privacy remain an issue in IoT due to its openness. Smart devices are manufactured by different vendors posing heterogeneity issue. Contributions are sought from different stakeholders.

In this chapter, an overview of key enabling technologies of IoT with focus on middleware technology is illustrated. But most of the middleware solutions are not generic. Middleware solutions must be aware of reliability and interoperability. A few middleware solutions provide context awareness. Further research can be directed towards interoperability, ambient intelligence, and context-awareness.

REFERENCES

1. J. Manyika, M. Chui, J. Bughin, R. Dobbs, P. Bisson, and A. Marrs, *Disruptive technologies: Advances that will transform life, business, and the global economy*, vol. 180. McKinsey Global Institute, San Francisco, CA, 2013.
2. L. Atzori, A. Iera, and G. Morabito, "SIoT: giving a social structure to the internet of things," *IEEE Commun. Lett.*, vol. 15, no. 11, pp. 1193–1195, 2011, DOI: 10.1109/LCOMM.2011.090911.111340.
3. R. Khan, S. U. Khan, R. Zaheer, and S. Khan, "Future internet: the internet of things architecture, possible applications and key challenges," in *2012 10th International Conference on Frontiers of Information Technology (FIT)*, 2012, pp. 257–260.
4. Z. Yang, Y. Yue, Y. Yang, Y. Peng, X. Wang, and W. Liu, "Study and application on the architecture and key technologies for IOT," in *International Conference on Multimedia Technology (ICMT'2011)*, 2011, pp. 747–751.
5. A. Al-Fuqaha, M. Guizani, M. Mohammadi, M. Aledhari, and M. Ayyash, "Internet of things: a survey on enabling technologies, protocols, and applications," *IEEE Commun. Surv. Tutor.*, vol. 17, no. 4, pp. 2347–2376, 2015, DOI: 10.1109/COMST.2015.2444095.
6. H. Ning and Z. Wang, "Future internet of things architecture: like mankind neural system or social organization framework?," *IEEE Commun. Lett.*, vol. 15, no. 4, pp. 461–463, 2011, DOI: 10.1109/LCOMM.2011.022411.110120.
7. A. Giri, S. Dutta, and S. Neogy, "Enabling agricultural automation to optimize utilization of water, fertilizer and insecticides by implementing Internet of Things (IoT)," in *International Conference on Information Technology (InCITe)—The Next Generation IT Summit on the Theme—Internet of Things: Connect Your Worlds*, 2016, pp. 125–131, DOI: 10.1109/INCITE.2016.7857603.
8. X. Jia, Q. Feng, T. Fan, and Q. Lei, "RFID technology and its applications in Internet of Things (IoT)," in *2nd International Conference on Consumer Electronics, Communications and Networks (CECNet''2012)*, 2012, pp. 1282–1285, DOI: 10.1109/CECNet.2012.6201508.
9. G. M. Lee, N. Crespi, J. K. Choi, and M. Boussard, "Internet of things," in *Evolution of Telecommunication Services*, Springer, Berlin, Heidelberg, 2013, pp. 257–282.
10. L. Da Xu, W. He, and S. Li, "Internet of things in industries: a survey," *IEEE Trans. Ind. Inform.*, vol. 10, no. 4, pp. 2233–2243, 2014.
11. E. Ferro and F. Potorti, "Bluetooth and Wi-Fi wireless protocols: a survey and a comparison," *IEEE Wirel. Commun.*, vol. 12, no. 1, pp. 12–26, 2005.
12. N. Kushalnagar, G. Montenegro, and C. Schumacher, "IPv6 over low-power wireless personal area networks (6LoWPANs): overview, assumptions, problem statement, and goals," *RFC 4919 (Informational), Internet Engineering Task Force*, 2007.
13. L. Atzori, A. Iera, and G. Morabito, "The internet of things: a survey," *Comput. Networks*, vol. 54, no. 15, pp. 2787–2805, 2010, DOI: http://dx.doi.org/10.1016/j.comnet.2010.05.010.
14. M. A. Razzaque, M. Milojevic-Jevric, A. Palade, and S. Clarke, "Middleware for internet of things: a survey," *IEEE Internet Things J.*, vol. 3, no. 1, pp. 70–95, 2016.
15. S. Bandyopadhyay, M. Sengupta, S. Maiti, and S. Dutta, "Role of middleware for internet of things: a study," *Int. J. Comput. Sci. Eng. Surv.*, vol. 2, no. 3, pp. 94–105, 2011.

16. R. Want, "Near field communication," *IEEE Pervas. Comput.*, vol. 10, no. 3, pp. 4–7, 2011.
17. Z. Shelby, K. Hartke, and C. Bormann, "The constrained application protocol (CoAP)," RFC—Proposed Standard (June 2014; Errata), 2014.
18. C. Bormann, A. P. Castellani, and Z. Shelby, "COAP: an application protocol for billions of tiny internet nodes," *IEEE Internet Comput.*, vol. 16, no. 2, pp. 62–67, 2012.
19. M. Eisenhauer, P. Rosengren, and P. Antolin, "Hydra: a development platform for integrating wireless devices and sensors into ambient intelligence systems," in *The Internet of Things*, Springer, Berlin Heidelberg, 2010, pp. 367–373.
20. P. R. Pietzuch, "*Hermes: A scalable event-based middleware*," University of Cambridge, Computer Laboratory, Technical Report No. 590, 2004.
21. P. Saint-Andre, "Extensible messaging and presence protocol (XMPP): Core," *Internet Eng. Task Force (IETF)*, Fremont, CA, Req. Comments 6120, 2011.
22. E. Avilés-López and J. A. García-Macías, "TinySOA: a service-oriented architecture for wireless sensor networks," *Serv. Oriented Comput. Appl.*, vol. 3, no. 2, pp. 99–108, 2009.
23. L. Baresi, S. Guinea, and P. Saeedi, "Achieving self-adaptation through dynamic group management," in *Assurances for Self-Adaptive Systems*, Springer, Berlin Heidelberg, 2013, pp. 214–239.
24. A. Giri, S. Dutta, S. Neogy, K. Dahal, and Z. Pervez, "Internet of things (IoT): A survey on architecture, enabling technologies, applications and challenges," in *ACM International Conference Proceeding Sereis*, 2017, DOI: 10.1145/3109761.3109768.
25. S. Wang, Z. Zhang, Z. Ye, X. Wang, X. Lin, and S. Chen, "Application of environmental internet of things on water quality management of urban scenic river," *Int. J. Sustain. Dev. World Ecol.*, vol. 20, no. 3, pp. 216–222, 2013.
26. O. Vermesan *et al.*, "Internet of things strategic research roadmap," *Internet Things-Global Technol. Soc. Trends*, vol. 1, pp. 9–52, 2011.
27. K. E. Kjær, "A survey of context-aware middleware," in *Proceedings of the 25th Conference on IASTED International Multi-Conference: Software Engineering*, 2007, pp. 148–155.
28. A. Boulis, C.-C. Han, R. Shea, and M. B. Srivastava, "SensorWare: programming sensor networks beyond code update and querying," *Pervasive Mob. Comput.*, vol. 3, no. 4, pp. 386–412, 2007.
29. G. Acampora, D. J. Cook, P. Rashidi, and A. V Vasilakos, "A survey on ambient intelligence in healthcare," *Proc. IEEE*, vol. 101, no. 12, pp. 2470–2494, 2013.
30. M. S. Shahamabadi, B. B. M. Ali, P. Varahram, and A. J. Jara, "A network mobility solution based on 6LoWPAN hospital wireless sensor network (NEMO-HWSN)," in *2013 Seventh International Conference on Innovative Mobile and Internet Services in Ubiquitous Computing (IMIS),* 2013, pp. 433–438.
31. P. Costa *et al.*, "The RUNES middleware for networked embedded systems and its application in a disaster management scenario," in *Pervasive Computing and Communications, 2007. PerCom'07. Fifth Annual IEEE International Conference on*, 2007, pp. 69–78.
32. M. Musolesi, C. Mascolo, and S. Hailes, "Emma: epidemic messaging middleware for ad hoc networks," *Pers. Ubiquitous Comput.*, vol. 10, no. 1, pp. 28–36, 2006.
33. T. Sivaharan, G. Blair, and G. Coulson, "Green: a configurable and re-configurable publish-subscribe middleware for pervasive computing," in *OTM Confederated International Conferences "On the Move to Meaningful Internet Systems*, 2005, pp. 732–749.
34. V. Tsiatsis, and A. Gluhak, "The SENSEI real world internet architecture," April 2010, DOI: 10.3233/978-1-60750-539-6-247.

35. C. M. Kirsch, M. A. A. Sanvido, and T. A. Henzinger, "A programmable microkernel for real-time systems," in *Proceedings of the 1st ACM/USENIX International Conference on Virtual Execution Environments*, 2005, pp. 35–45.
36. K. Hong *et al.*, "TinyVM: an energy-efficient execution infrastructure for sensor networks," *Softw. Pract. Exp.*, vol. 42, no. 10, pp. 1193–1209, 2012.
37. P. Kang, C. Borcea, G. Xu, A. Saxena, U. Kremer, and L. Iftode, "Smart messages: a distributed computing platform for networks of embedded systems," *Comput. J.*, vol. 47, no. 4, pp. 475–494, 2004.
38. C.-L. Fok, G.-C. Roman, and C. Lu, "Agilla: a mobile agent middleware for self-adaptive wireless sensor networks," *ACM Trans. Auton. Adapt. Syst.*, vol. 4, no. 3, p. 16, 2009.
39. V. Terziyan, O. Kaykova, and D. Zhovtobryukh, "Ubiroad: semantic middleware for context-aware smart road environments," in *2010 Fifth International Conference on Internet and Web Applications and Services (ICIW)*, 2010, pp. 295–302.
40. A. L. Murphy, G. Pietro Picco, and G.-C. Roman, "Lime: a middleware for physical and logical mobility," in *21st International Conference on Distributed Computing Systems-2001*, 2001, pp. 524–533.
41. R. de C. A. Lima, N. S. Rosa, and I. R. L. Marques, "TS-Mid: middleware for wireless sensor networks based on tuple space," in *22nd International Conference on Advanced Information Networking and Applications-Workshops (AINAW 2008)*, 2008, pp. 886–891.
42. P. Bonnet, J. Gehrke, and P. Seshadri, "Towards sensor database systems," in *International Conference on Mobile Data Management*, 2001, pp. 3–14.
43. T. Hasiotis, G. Alyfantis, V. Tsetsos, O. Sekkas, and S. Hadjiefthymiades, "Sensation: a middleware integration platform for pervasive applications in wireless sensor networks," in *Proceedings of the Second European Workshop on Wireless Sensor Networks 2005*, 2005, pp. 366–377.
44. S. R. Madden, M. J. Franklin, J. M. Hellerstein, and W. Hong, "TinyDB: an acquisitional query processing system for sensor networks," *ACM Trans. Database Syst.*, vol. 30, no. 1, pp. 122–173, 2005.
45. P. Andreou, D. Zeinalipour-Yazti, M. Vassiliadou, P. K. Chrysanthis, and G. Samaras, "Kspot: effectively monitoring the k most important events in a wireless sensor network," in *IEEE 25th International Conference on Data Engineering, 2009 (ICDE'09)*, 2009, pp. 1503–1506.
46. W. B. Heinzelman, A. L. Murphy, H. S. Carvalho, and M. A. Perillo, "Middleware to support sensor network applications," *IEEE Networks*, vol. 18, no. 1, pp. 6–14, 2004.
47. H. Alex, M. Kumar, and B. Shirazi, "MidFusion: an adaptive middleware for information fusion in sensor network applications," *Inf. Fusion*, vol. 9, no. 3, pp. 332–343, 2008.
48. Q. Han and N. Venkatasubramanian, "Autosec: an integrated middleware framework for dynamic service brokering," *IEEE Distrib. Syst. Online*, vol. 2, no. 7, pp. 22–31, 2001.
49. N. Bui and M. Zorzi, "Health care applications: a solution based on the internet of things," in *Proceedings of the 4th International Symposium on Applied Sciences in Biomedical and Communication Technologies*, 2011, p. 131.
50. A.-M. Rahmani *et al.*, "Smart e-health gateway: bringing intelligence to internet-of-things based ubiquitous healthcare systems," in *Consumer Communications and Networking Conference (CCNC), 2015 12th Annual IEEE*, 2015, pp. 826–834.
51. A. Dohr, R. Modre-Opsrian, M. Drobics, D. Hayn, and G. Schreier, "The internet of things for ambient assisted living," in *2010 Seventh International Conference on Information Technology: New Generations (ITNG)*, 2010, pp. 804–809.
52. M. Jude, "IBM: Working Towards a Smarter Connected Home. Internet." http://docs.caba.org/documents/IBM-Smart-Cloud-Home-SPIE2012.pdf, 2014.

53. L. Atzori, A. Iera, G. Morabito, and M. Nitti, "The social internet of things (SIoT)–when social networks meet the internet of things: concept, architecture and network characterization," *Comput. Networks*, vol. 56, no. 16, pp. 3594–3608, 2012.
54. J. Kleinberg, "The convergence of social and technological networks," *Commun. ACM*, vol. 51, no. 11, pp. 66–72, 2008.
55. P. Semmelhack, *Social machines: how to develop connected products that change customers' lives*. John Wiley & Sons, Hoboken, New Jersey, 2013.
56. A. M. Vilamovska, E. Hattziandreu, R. Schindler, C. Van Oranje, H. De Vries, and J. Krapelse, "RFID application in healthcare–scoping and identifying areas for RFID deployment in healthcare delivery," *RAND Eur.* Feb., 2009.
57. L. Lei-Hong, H. Yue-Shan, and W. Xiao-Ming, "A community health service architecture based on the internet of things on health-care," in *World Congress on Medical Physics and Biomedical Engineering*, May 26–31, 2012, *Beijing, China*, 2013, pp. 1317–1320.
58. J. Granjal, E. Monteiro, and J. S. Silva, "Security for the internet of things: a survey of existing protocols and open research issues," *IEEE Commun. Surv. Tutor.*, vol. 17, no. 3, pp. 1294–1312, 2015.
59. AllJoyn Framework, www.openconnectivity.org/search/AllJoyn,
60. S. M. R. Islam, D. Kwak, M. D. H. Kabir, M. Hossain, and K.-S. Kwak, "The internet of things for health care: a comprehensive survey," *IEEE Access*, vol. 3, pp. 678–708, 2015.
61. H. A. Khattak, M. Ruta, and E. Di Sciascio, "CoAP-based healthcare sensor networks: a survey," in *2014 11th International Bhurban Conference on Applied Sciences and Technology (IBCAST)*, 2014, pp. 499–503.
62. E. C. Larson, M. Goel, G. Boriello, S. Heltshe, M. Rosenfeld, and S. N. Patel, "SpiroSmart: using a microphone to measure lung function on a mobile phone," in *Proceedings of the 2012 ACM Conference on Ubiquitous Computing*, 2012, pp. 280–289.
63. R. S. H. Istepanian, S. Hu, N. Y. Philip, and A. Sungoor, "The potential of Internet of m-health Things 'm-IoT' for non-invasive glucose level sensing," in *Engineering in Medicine and Biology Society, EMBC, 2011 Annual International Conference of the IEEE*, 2011, pp. 5264–5266.
64. P. K. Dash, "Electrocardiogram monitoring," *Indian J. Anaesth*, vol. 46, no. 4, pp. 251–260, 2002.
65. L. Yang, Y. Ge, W. Li, W. Rao, and W. Shen, "A home mobile healthcare system for wheelchair users," in *Computer Supported Cooperative Work in Design (CSCWD), Proceedings of the 2014 IEEE 18th International Conference on*, 2014, pp. 609–614.
66. "Dr. Hawking's Connected Wheelchair Project, Intel, Health and Human Services Connectivity, 2014.
67. D. Miorandi, S. Sicari, F. De Pellegrini, and I. Chlamtac, "Internet of things: vision, applications and research challenges," *Ad Hoc Networks*, vol. 10, no. 7, pp. 1497–1516, 2012.
68. A. Zanella, N. Bui, A. Castellani, L. Vangelista, and M. Zorzi, "Internet of things for smart cities," *IEEE Internet Things J.*, vol. 1, no. 1, pp. 22–32, 2014.
69. T. Nam and T. A. Pardo, "Conceptualizing smart city with dimensions of technology, people, and institutions," in *Proceedings of the 12th Annual International Digital Government Research Conference: Digital Government Innovation in Challenging Times*, 2011, pp. 282–291.
70. T. Bakıcı, E. Almirall, and J. Wareham, "A smart city initiative: the case of Barcelona," *J. Knowl. Econ.*, vol. 4, no. 2, pp. 135–148, 2013.
71. H. Wang and W. He, "A reservation-based smart parking system," in *Computer Communications Workshops (INFOCOM WKSHPS), 2011 IEEE Conference on*, 2011, pp. 690–695.

72. D. Bonino *et al.*, "Almanac: Internet of things for smart cities," in *Future Internet of Things and Cloud (FiCloud), 2015 3rd International Conference on*, 2015, pp. 309–316.
73. M. Yun and B. Yuxin, "Research on the architecture and key technology of internet of things (IoT) applied on smart grid," in *Advances in Energy Engineering (ICAEE), 2010 International Conference on*, 2010, pp. 69–72.
74. E. Qin, Y. Long, C. Zhang, and L. Huang, "Cloud computing and the internet of things: technology innovation in automobile service," in *International Conference on Human Interface and the Management of Information*, 2013, pp. 173–180.
75. S. Raza, H. Shafagh, K. Hewage, R. Hummen, and T. Voigt, "Lithe: lightweight secure CoAP for the internet of things," *IEEE Sens. J.*, vol. 13, no. 10, pp. 3711–3720, 2013.
76. S. Sicari, A. Rizzardi, L. A. Grieco, and A. Coen-Porisini, "Security, privacy and trust in internet of things: the road ahead," *Comput. Networks*, vol. 76, pp. 146–164, 2015.
77. I. T. S. Headquarters, "e-Japan strategy," Retrieved June, vol. 22, p. 2004, 2001.
78. G. Baldini *et al.*, "A cognitive framework for realizing and exploiting the internet of things concept," in *27th WWRF Meeting*, Dusseldorf, 2011.
79. "IoT6 European research project, Deliverable D1.5:", www.iot6.eu/handbook
80. "IoTivity.", www.iotivity.in
81. P. Costa, G. Coulson, C. Mascolo, G. P. Picco, and S. Zachariadis, "The RUNES middleware: a reconfigurable component-based approach to networked embedded systems," in *2005 IEEE 16th International Symposium on Personal, Indoor and Mobile Radio Communications*, vol. 2, pp. 806–810, 2005, DOI: 10.1109/PIMRC.2005.1651554.
82. X. Wu, X. Zhu, G.-Q. Wu, and W. Ding, "Data mining with big data," *IEEE Trans. Knowl. Data Eng.*, vol. 26, no. 1, pp. 97–107, 2014.
83. C.-W. Tsai, C.-F. Lai, M.-C. Chiang, and L. T. Yang, "Data mining for internet of things: a survey," *IEEE Commun. Surv. Tutor.*, vol. 16, no. 1, pp. 77–97, 2014.
84. M. Moshtaghi *et al.*, "Streaming analysis in wireless sensor networks," *Wirel. Commun. Mob. Comput.*, vol. 14, no. 9, pp. 905–921, 2014.
85. A. Bialecki, M. Cafarella, D. Cutting, and O. O'Malley, "Hadoop: a framework for running applications on large clusters built of commodity hardware," *Wiki*, vol. 11, 2005, http//lucene.apache.org/hadoop.
86. P. G. Brown, "Overview of SciDB: large scale array storage, processing and analysis," in *Proceedings of the 2010 ACM SIGMOD International Conference on Management of data*, 2010, pp. 963–968.

5 IoT Service Platform

Prachet Bhuyan and Abhishek Ray
School of Computer Engineering, KIIT
Deemed to be University, Odisha, India

CONTENTS

5.1 Introduction to IoT Platform 91
5.2 Referential IoT Platforms 92
5.3 Challenges of the IoT Platform 94
5.3.1 Adoption and Application of an IoT Platform 94
5.3.2 Scalability of IoT Platform 94
5.3.3 Hardware and Software Compatibility 94
5.3.4 The Various Vendors Providing IoT Service Platform 94
5.3.5 Standard Protocols of IoT Platform 95
5.3.6 Security and Privacy in IoT Platform 95
5.3.7 IoT Platform Challenges—An Industry Perspective 95
5.3.7.1 Scalability 95
5.3.7.2 Interoperability 96
5.3.7.3 Design-Based Challenge 96
5.3.7.4 Security 96
5.3.7.5 Technology Infrastructure 96
5.4 Conclusion and Future of IoT Platform 96
References 97

5.1 INTRODUCTION TO IoT PLATFORM

The challenges of the new smart world, the requirement of the emerging connectivity of billions of devices, and real-time systems demand an integrated Internet of Things (IoT) platform as a necessity. The IoT platform itself goes beyond the description of IoT devices. A platform is an echo system to nurture the demand for cooperating and integrated services of IoT-based applications. It is an end-to-end approach to enable all the possible IoT device connections, its configuration in the application areas, the relevant data collected from the sensors, its transport, and storage in the cloud, further predicting the features or patterns processing the sensed data from IoT sensors.

As an example of a classic case study, the smart society charter for IoT architecture gives principles and guidelines for the city of Eindhoven for the smart society. Digital online technologies got integrated to improve the quality of life. The simultaneous growth of digital, analog, virtual, physical, online, and offline benefits the end users. The advancement of technologies with IoT, the voluminous and veracity of data being collected, and IoT along data-oriented services is going to revolutionize in the society. Various stakeholder's concerns and their needs of the general public, a business model with innovations, and sustainability become the craftsmanship of cooperating rules and regulations of the emerging IoT-related technologies collectively. All the views and expectations of stakeholders, like small and medium enterprises, corporate houses, new start-up ventures, IoT enthusiasts, research organizations and institutions, government, and private partners, etc., collectively contribute to the strength and build a robust data-driven framework, with related technologies, along with the integrated IoT platform [4].

5.2 REFERENTIAL IoT PLATFORMS

Addressing the need for an integrated approach to support the context and environment under which all these IoT-based applications can work efficiently, a proposed blueprint of a referential IoT platform architecture is shown in Figure 5.1.

The proposed IoT platform architecture shown in Figure 5.1 is a generic referential purpose only. As far as our knowledge goes, various IoT platforms have already been active in the commercial market. Each IoT platform is trying to resolve and integrate as the IoT-based applications are adopted across various domains. Hence, the challenge becomes more to fulfill the demand of various domains.

In Figure 5.1, three separate modules, 1, 2, and 3, cater to the need of integrating the IoT-based applications. This proposed referential IoT platform can be fine-tuned for various domains or sectors by adopting suitable changes.

Module 1 (IoT devices with sensors, computing and communication) related to IoT devices with sensors, computing, and communication modules, as shown in Figure 5.1, can be packaged as one device with IoT sensors and the computing module acting as one node. Various such nodes can be deployed across the actual place or field of applications. Data can be sensed, and through different mediums of communication using a wired or a wireless medium, it can be sent or received.

Module 2 (Fast and reliable communication medium) deals with the transportation of data to the required destination server, which may be in cloud or cluster or larger-capacity computing nodes. It may use a wired-like fiber, or wireless, or combination of both, forming a hybrid communication channel also. The challenge here is to select a cost-effective medium of transport.

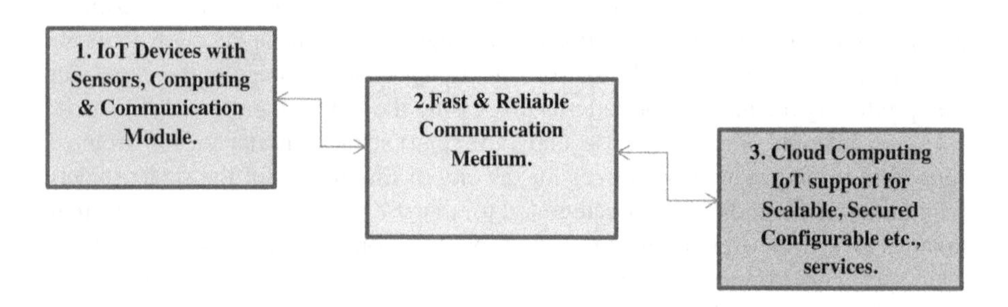

FIGURE 5.1 Proposed referential IoT platform architecture.

For example, short-range IoT devices in healthcare may not be the same as smart manufacturing, or agriculture. The distance, type, volume, and many similar factors will affect the choice of communication medium.

Module 3 (Cloud computing IoT support for scalable, secured configurable, etc.,) deals with the collection of data from various endpoints or sensors or integrated IoT devices at a central or distributed computing nodes in cloud or cluster, or larger-capacity computing nodes or edge computing-based devices. Here again, our objective is to be flexible to cater to the need of various IoT applications across domains. Some applications only want to capture raw data, and some applications want to process relevant meaningful data for pattern analysis and predict various safe or unsafe patterns for making some important decisions for the process. It may be in a supply chain cycle, may be in a smart fracturing process, in healthcare, etc.

The three modules, 1, 2, and 3, primarily are flexible in adopting technologies, related to IoT and data analytic, artificial intelligence, machine learning, deep learning, and so on for various types of data model predictions, and secured, reliable communication of data, and a responsive server may be in cloud or cluster or local computing nodes. Aside from all these, the factors for scalable, secured, and reliable communication can be fine-tuned based on the varied domain of applications.

There are even various other open-source referential IoT platforms proposed, which also cater to the need of niche technology and other allied areas of IoT. Seoyeon Kim *et al.* surveyed some IoT platforms. They described the main features of these IoT platforms and looked at whether they support artificial intelligence (AI) technology. Table 5.1 shows the result of all surveyed IoT platforms as done by Seoyeon Kim *et al.* [3].

Seoyeon Kim *et al.* had emphasized finding out the usage of AI technologies in some open-source IoT platforms, and some related IoT platforms were discussed as well. Table 5.1 gives a brief description of all the compared IoT platforms, mentioning whether the support of AI is there or missing. Hence this also throws light that the IoT platform can vary for the various domains of applications [3].

TABLE 5.1
IoT Platforms Supporting AI [3]

IoT Platform	ML (Machine Learning) Library	Description
Thingsboard [1]	Third-party	Examples: Apache Spark, Kafka
Kaa [2]	Support	Custom project
Node-Red [3]	Support	Need to install node-red-contrib-machine-learning
Ptolemy [4]	Support	Particle filtering, model predictive control, hidden Markov models, and multiple tools for statistical analysis
MIS-IoT [5]	Support	Type of modular machine learning algorithm

5.3 CHALLENGES OF THE IoT PLATFORM

IoT is a buzz word in the current market scenario and is in demand as a niche technology, along with artificial intelligence, data analytics, cloud, and so on, and its wide acceptable domain of applications, ranging from agriculture, healthcare, industrial manufacturing, defense, smart city, and transportation. This makes IoT a challenging field altogether, and it needs wide cooperation to integrate the solution, keeping in balance the expectations of customers, government policies, vendors, technology partners, device manufacturers, and social acceptance. There will be widely varied challenges for IoT platform, such as the following areas.

5.3.1 Adoption and Application of an IoT Platform

The more IoT awareness is spreading, and its acceptance in various sectors, makes the evolution process of IoT platform a huge challenge. The various demands for different applications bring in more integration issues for the IoT platform. One common IoT platform cannot guarantee to be the best fit for all applications across domains.

5.3.2 Scalability of IoT Platform

The IoT platform needs to address the various IoT devices and the computing demand of IoT-based applications. It needs to be scaled up for an on-demand approach as per the various IoT device's needs, their communication, and the computing requirements. Communication itself is a big contributor to scalable IoT device's integration. In the era of telecommunication boom, optical fiber, wireless, and mobile communication, along with other short-range communication mediums like Bluetooth, ZigBee, RFID, etc., having standard protocols as per IEEE, ETSI, and ITU norms, ensures that it gets synchronized along with the IoT platform end-to-end from the devices to the IoT cloud nodes.

5.3.3 Hardware and Software Compatibility

The devices and their composition may vary from applications to applications among various domains. The standardization of IoT devices hence may not be possible as a single device across all domains, and it's a long way to settle for common standardized devices. It needs to be customized as per its application, considering sensors, memory, communication, and computing demand, etc., to give its best acceptable performance.

5.3.4 The Various Vendors Providing IoT Service Platform

As the IoT applications gain wider acceptance in different domains, like agriculture, healthcare, weather, defense, image processing, transportation, and industrial manufacturing, etc., the complete know-how with a single vendor will always be a challenge, as the IoT technology can be blended with various other technologies, such as both hardware, software, and communication in particular, to suit and deliver the exact requirements. Hence the awareness and expertise about the complete know-how of the IoT platform remains a challenge and can be resolved with strategic alliances and collaborations.

5.3.5 Standard Protocols of IoT Platform

As the IoT platform is being offered by various organizations, it has yet to see a worldwide standardization of the platform. Through various technologies, like cloud, mobile communication, and IoT custom devices, there is a wide scope and challenges for standardization of the IoT platform. Though there are some best practices, yet the challenge to have a common standard IoT platform is a bit far distance.

5.3.6 Security and Privacy in IoT Platform

This is one of the important concerns, as millions and billions of devices that communicates with each device, and security and privacy need to be protected. Due to malicious attacks, hacking, privacy violation of personal data, and so on, it becomes mandatory to have regular checks and monitoring, and around the clock, the activities of all IoT devices for communications and transactions. Robust security and privacy mechanism is the need of the hour.

There have been specific efforts been put by Z. Berkay Celik *et al.* to study privacy and security issues in IoT that require program-analysis techniques with an emphasis on identified attacks against these systems and defenses implemented so far. Based on a study of five IoT programming platforms, they identified the key insights that result from research efforts in both the program analysis and security communities, and they relate the efficacy of program-analysis techniques to security and privacy issues [2].

5.3.7 IoT Platform Challenges—An Industry Perspective

To complement the above-mentioned challenges from subsections 5.3.1 to 5.3.6, here are some of the industrial experiences of IoT platform challenges, and its adaptability are as follows.

The IoT has been facing many areas like information technology, healthcare, data analytics, and agriculture. The main focus is on protecting privacy, as it is the primary reason for other challenges, including government participation and its rules, regulations, and policies [6].

5.3.7.1 Scalability

Billions of internet-enabled devices get connected in a huge network, and large volumes of data need to be processed. The system that stores and analyses the data from these IoT devices needs to be scalable. At the present time, the era of IoT evolution and everyday objects is connected via the internet. The raw data obtained from these devices needs Big Data analytics and cloud storage for the interpretation of useful data [6].

5.3.7.2 Interoperability

Technological standards in most areas are still fragmented. These technologies need to be converged, which would help us in establishing a common framework and the standard for the IoT devices. As the standardization process is still lacking, the interoperability of IoT with legacy devices should be considered critical. This lack of interoperability is preventing us from moving toward a common standard platform [6].

5.3.7.3 Design-Based Challenge

With the development in technology design, challenges are increasing at a faster rate. There have been issues regarding design, like limited computation power, limited energy, and limited memory, which need to be sorted out for better-packaged devices [6].

The biggest challenge is to decide to integrate from product to cloud or, with various cooperation, with prospects to vendors. The options for regular users of the IoT platform also face issues like the devices to be designed, its deployment with integration, and ready-made IoT service platform for integration [7].

5.3.7.4 Security

As per another renowned company's concern, exchanging data over the internet always has security issues. Hacking is an international industry, producing frequent announcements of security breaches. Putting data online, particularly data related to critical equipment, may seem dangerous. IoT security assessments consider security from multiple aspects, which are as follows [1].

5.3.7.5 Technology Infrastructure

Technology infrastructure may prohibit exchanging of data to IoT platform due to security violations. *Communications infrastructure:* Using a cellular gateway to connect IoT instruments sounds great, but users don't get phone reception at some remote sites. Building infrastructure would be too costly. *Immaturity of IoT standards:* There was hesitance in leading the way forward in the IoT standards scenario. Waiting to see which standard or protocol would win results in delayed IoT investments. The Open Connectivity Foundation joined the Open Interconnect Consortium and is pushing a united protocol. The Institute of Electrical and Electronics Engineers (IEEE) published its draft P2413 standard for IoT architecture, creating a universal language for IoT that would greatly reduce the effort required to share data among competing platforms. Regardless of which platform is chosen, users will soon be able to share data across all IoT devices and platforms. *Procuring IoT:* Implementing IoT often involves procuring devices and services that don't have IoT in their names, such as instrumentation, communication networks, storage, and data management consultants [1].

5.4 CONCLUSION AND FUTURE OF IoT PLATFORM

IoT undoubtedly has evolved as one of the necessary technologies widely accepted across various domains. Society benefits from this emerging IoT technology. Due to its wide range of applications in varied domains, the challenges of the IoT platform increase, and they need a standard protocol-based approach to address it. Various

IoT platforms have envisaged these evolving issues from time to time. In this chapter, there has been an effort to see the IoT platform in a generic reference way, and the variations of the IoT platform that can occur are also discussed. The IoT platform can be complimented, along with other technologies, like AI, Block Chain, Cloud Computing, etc. The scalability, integration, security, and interoperability, along with the fundamental capabilities of the IoT platform, make it a robust and flexible program to support IoT-based applications with demanding features. The scope of its acceptance is even becoming more with industrial IoT emergence, with technologies like Edge Computing, Block Chain, Machine Learning, Deep Learning, Big Data, etc., making it a sought-after game-changer in the commercial market across all fields. The IoT platform is a boon for improving the efficiency and productivity in various domains of applications.

REFERENCES

1. https://www.controleng.com/articles/six-iot-implementation-challenges-and-solutions/: Industry experience sharing of IoT Platform.
2. Z. Berkay Celik, Earlence Fernandes, Eric Pauley, Gang Tan, and Patrick McDaniel. Program Analysis of Commodity IoT Applications for Security and Privacy: Challenges and Opportunities. *ACM Computing Surveys*, vol. 52, no. 4, Article 74 (August 2019), 30 pp. https://doi.org/10.1145/3333501
3. S. Kim, J. Park, J. Jeong, Y.S. Yun, S. Eun, and J. Jung. Survey of IoT Platforms Supporting Artificial Intelligence. In Proceedings of International Conference on Research in Adaptive and Convergent Systems, Chongqing, China, September 24–27, 2019 (RACS'19), 2 pages. https://doi.org/10.1145/3338840.3355694
4. https://datasetsearch.research.google.com/search?query=IoT%20definition&docid=N2HaAgRD%2BQDaBL47AAAAAA%3D%3D.
5. Aras Can Onal, Omer Berat Sezer, Murat Ozbayoglu, and Erdogan Dogdu. MIS-IoT: Modular Intelligent Server Based Internet of Things Framework with Big Data and Machine Learning. In IEEE International Conference on Big Data, 10–13th December, 2018, DOI: 10.1109/BigData.2018.8622247
6. https://www.geeksforgeeks.org/challenges-in-world-of-iot/: Experience on IoT Platform.
7. https://www.happiestminds.com/blogs/the-challenge-of-iot-platforms/: Industry experience sharing of IoT Platform issues.

6 Software Integrated Framework Design for IoT-Based Applications

Sugyan Kumar Mishra and Anirban Sarkar
National Institute of Technology, Durgapur, India

CONTENTS

6.1 Introduction 99
6.2 Related Works 100
  6.2.1 Information Delivery Architecture 100
  6.2.2 Event-Driven Work-Flow-Based Requirements Analysis for IoT-Based Applications 101
  6.2.3 Quality Evaluation 104
6.3 Theoretical Foundation for Software Integrated Framework Design for IoT-Based Applications 105
  6.3.1 Information Architecture for IoT-Based Applications 106
  6.3.2 Evaluation of the Proposed Model Through a Case Study 107
6.4 Future Research Directions 108
  6.4.1 Enrichment of Hypergraph-Based Approach for IoT-Based SA 108
  6.4.2 Event-Driven Work Flow-Based Requirements Analysis for IoT-Based Applications 109
  6.4.3 Verification and Validation of the Service Composition Pattern 109
  6.4.4 Quality Evaluation Framework for IoT-Based SA 109
6.5 Conclusion 110
References 110

6.1 INTRODUCTION

According to the IEEE standard, software architecture (SA) presents a framework's outline to demonstrate, increment, and develop programming concentrated frameworks by determining computational elements or information stores of programming as structural segments and connectors [1]. SA models have been effectively used to adequately build mediums, just as intricate mechanical scale frameworks in the past [2]. The goal of SA for the Internet of Things (IoT) is to extract the complexities of heterogeneous equipment parts and system conventions [3, 4]. The existing research of SA can be misused to display, create, execute, and advance complex IoT frameworks that fulfil the ideal usefulness, just as the necessary quality [2, 3].

However, there is a requirement to design architectural solutions for the upcoming IoT-based frameworks [5–7].

Service-oriented architecture (SOA) was originated from a set of concepts and rules to web services (WS) standards [8]. SOA offers a suitable background for supporting the *connectivity*, *interoperability,* and *integration* in an IoT framework. Interoperability is one among the many difficulties in an IoT smart environment [9]. There exists a set of terms for explaining services, service constraints, and service strategies.

A business process task is nothing but a special unit implemented by a service [10, 11]. There exists a gap between the IoT framework and the context-aware business process modeling (BPM) [12]. Therefore, the integration is required to interact with its physical environment through service interfaces. Various business process notations (BPMN, BPEL, and JPDL) are existing on a scientific domain [11]. Architecture evaluations can be performed in one or several stages of a software development process.

The remaining sections are structured as follows. Section 6.2 presents various research work and also challenges related to discussed problem domains. Section 6.3 illustrates the proposed approach for designing of large-scale SOA (LSS). The proposed approach was also outlined through a suitable case study. Various research challenges are discussed for integration of semantic sensor network (SSN) ontology with the service. The scope of future work, such as integration of SSN ontology with hypergraph coloring (HC) approach, adaptation of IoT in the context of BPM, and finally, quality parameters to evaluate the SA, are discussed in Section 6.4. There should be a proper verification and validation tool for the service composition. Finally, the last section concludes with remarks.

6.2 RELATED WORKS

The outline of basic IoT infrastructure and the existing modeling approach for SOA is discussed in this section. The different challenges at the design and development levels of application software are also discussed. The SSN ontology is described, along with its limitations. An overview of the framework that connects the IoT infrastructure with BPM by integrating IoT data has been discussed in detail. A rigorous study is carried out in the context of IoT quality assurance.

6.2.1 Information Delivery Architecture

The basic IoT framework design is shown in Figure 6.1. The physical layer includes sensors, actuators, and hardware devices to gather real-world data. The sensor is a device, which sends the data to the processor. The middleware layer controls the interaction between the physical sensing layer and the application layer. The application layer sends orders to real-world objects over the web. It is utilized by the consumer [13].

There are several challenges, such as *configurability*, *interoperability*, and *manageability*. Configurability is the capacity of creating services for a large number of smart devices. Interoperability refers to the capacity of exchanging data among heterogeneous smart devices, utilizing heterogeneous communication protocols and related services. Manageability is the ability of managing the services on a huge

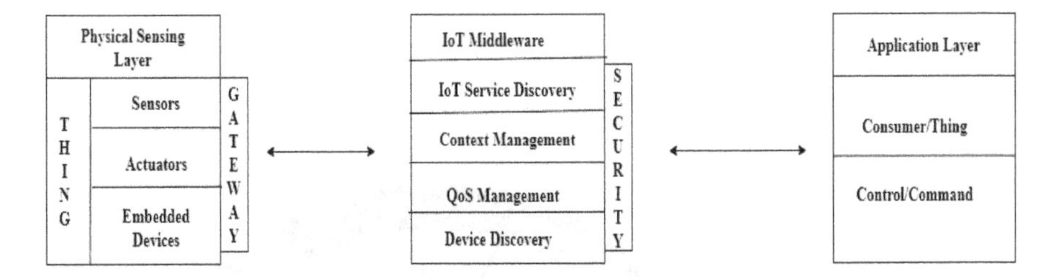

FIGURE 6.1 The IoT basic architecture [13].

number of smart devices remotely. SOA provides a suitable framework that supports the connectivity, interoperability, and integration in an IoT framework [13, 14].

Figure 6.2 displays three major elements of SOA, such as *service consumer* (SC), *service provider* (SP), and *registry* [15]. The service is published by the SP in the registry; the service request is sent by the SC. Finally, the *bind()* is established between the SC and the SP to use the service [16, 17]. Service composition is a way of designing complex service by considering the existing services [18].

Semantic approaches, i.e. ontologies, have been utilized as a strategy to solve the issues identified with large-scale heterogeneity. It is a collection of four major components, such as classes, relations, attributes, and individuals. The components of the ontology are classified into two kinds, such as Sensor, Observation, Sample, and Actuator (SOSA), and SSN [19]. Figure 6.3 describes the horizontal and vertical segmentation of SOSA [20]. The various classes and properties of SOSA are shown in Figure 6.4 [21].

6.2.2 Event-Driven Work-Flow-Based Requirements Analysis for IoT-Based Applications

Services are dynamically gathered in an opened situation, and these services may have various prerequisites. Hence, it turns out to be very challenging to determine if desired services should be a part of a workflow. Different workflow frameworks have been intended to work in several scientific domains like IoT. Vector Symbolic Architectures (VSAs) [22] are a lot of lossy dimensionality decrease systems that

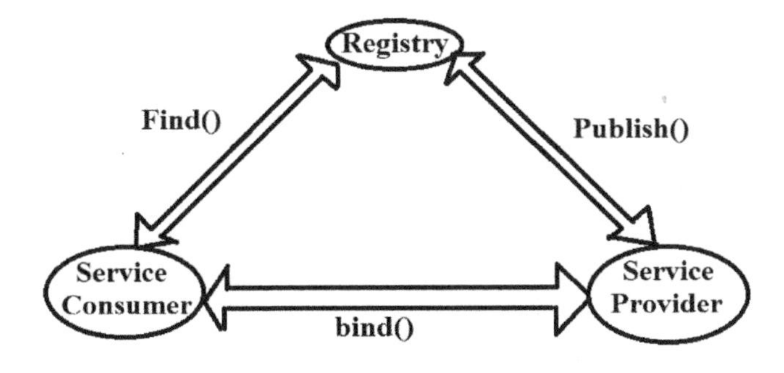

FIGURE 6.2 The basic structure of SOA.

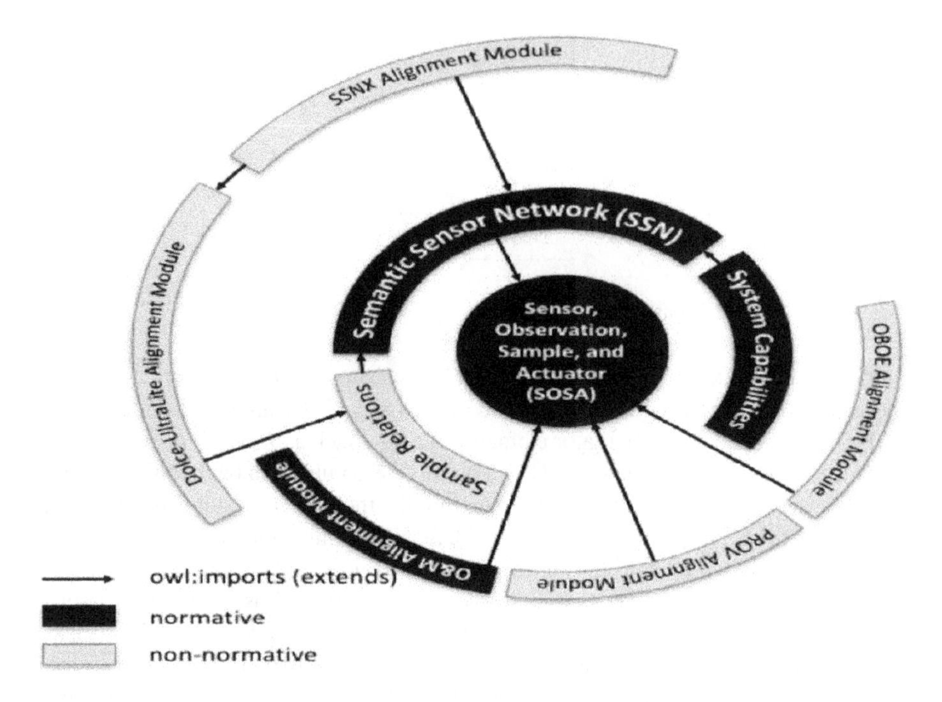

FIGURE 6.3 SOSA and its vertical and horizontal modules [20].

helps enormous volumes of information to be compacted into a fixed-size vector. Node-RED [23] is a graphical work process management system designed for IoT devices [24].

Schutte *et al.* [25] have presented LUCON, which considers data flows by considering the way that messages may be routed across services. It is designed for

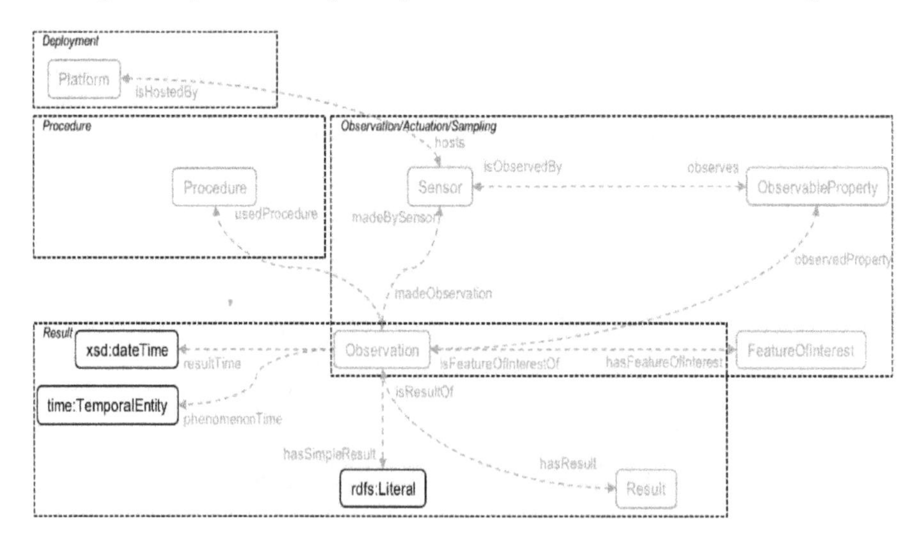

FIGURE 6.4 The properties of SOSA classes [21].

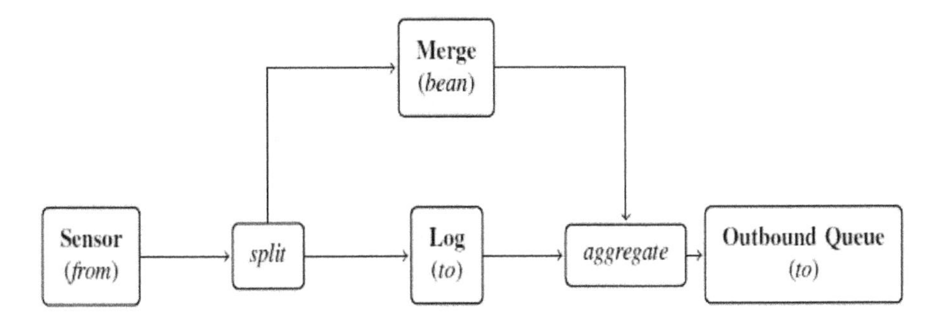

FIGURE 6.5 The path of the messages over services [25].

distributed systems. Figure 6.5 shows the path of messages over services. Pasquier *et al.* [26] have introduced CamFlow, an end-to-end information flow enforcement system. Various IoT concepts have been developed and designed for integrating with the BPM concepts [27–29]. Song *et al.* [18] have acquainted a system with diminish the hole and context-aware BPM. Figure 6.6 displays different components such as Business Process Models, Decision Models, Context Models, and Business Process Execution of the BPM ecosystem.

A framework is discussed in Figure 6.7 that associates the BPM environment with IoT infrastructure [18]. The use of different IoT techniques leads to an enormous number of IoT devices deployed. Sensor fusion is required for usability of

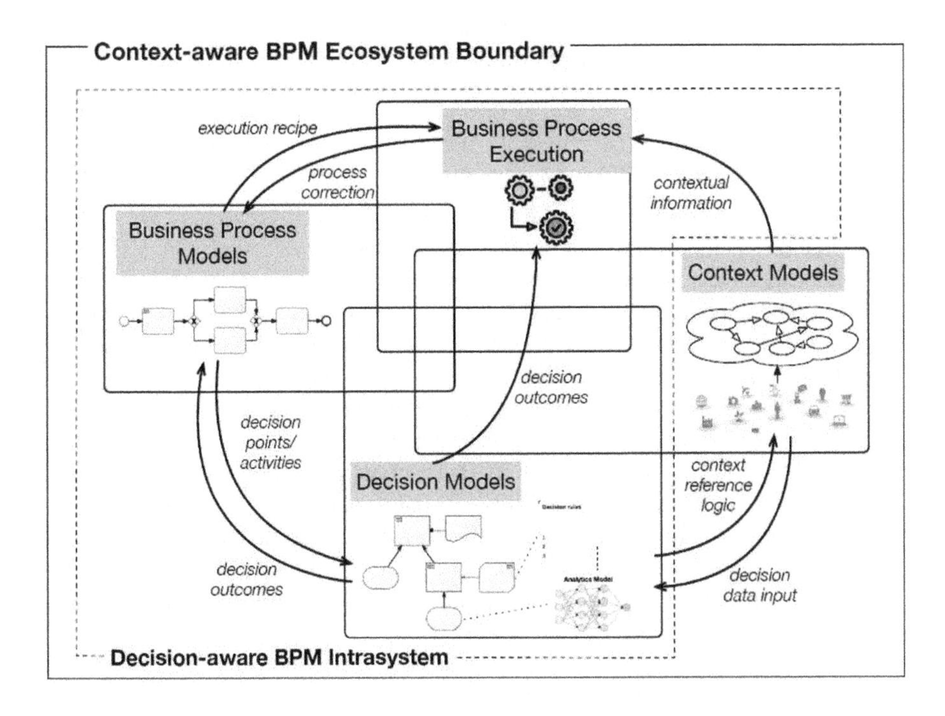

FIGURE 6.6 The context-aware BPM ecosystem [18].

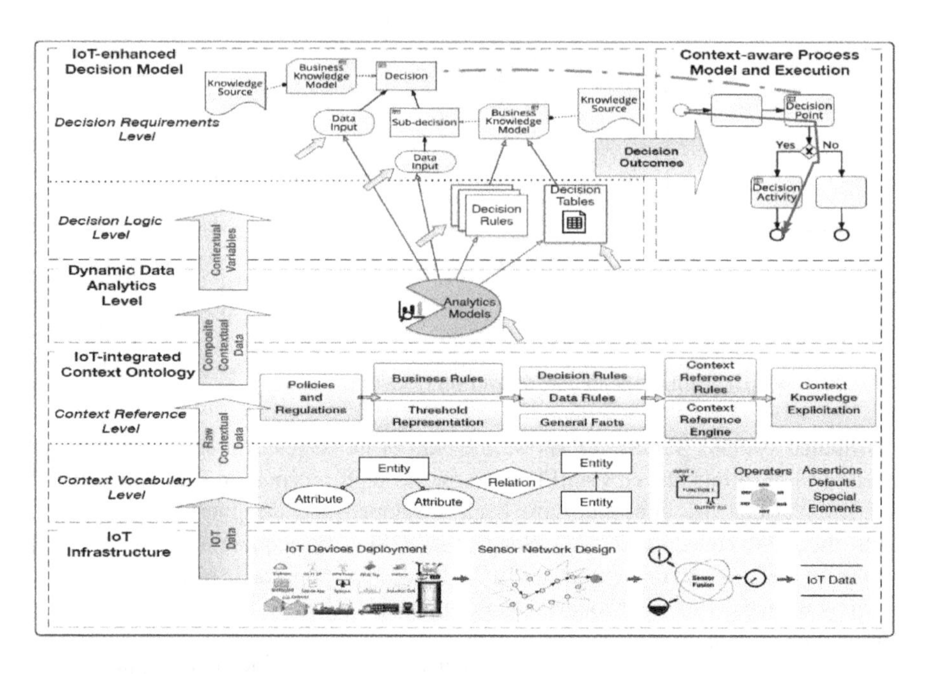

FIGURE 6.7 An integration of BPM ecosystem with IoT infrastructure [18].

IoT data. The IoT-integrated context model contains two levels. One is the context vocabulary level, which incorporates the structure of contextual elements. Another is the context reference level that depends on investigating business rule to plan setting reference rules.

6.2.3 Quality Evaluation

Various quality of service (QoS) factors are there in an IoT domain. The quality model is used as a parameter for service assessment and selection [30]. Kim characterized quality factors into various classes, for instance, *functionality*, *reliability*, *efficiency*, and *portability*. Liu *et al.* [31] have shown a QoS model of grey decision-making from the perspective on IoT worldwide foundation. A QoS model was introduced by Liu *et al.* [31] from the perspective on IoT global infrastructure.

The quality assurance procedures are a fundamental separation metric to select among the various cloud suppliers accessible currently. Thus, Zheng *et al.* [32] examined a quality model named CLOUDQUAL for cloud providers. It takes six notable quality measurements, such as reliability, availability, usability, security, elasticity, and responsiveness. However, Ahmad [33] has just focused on the reliability models out of six quality measurements. Karkouch *et al.* [34] proposed a model-driven architecture-based methodology for quality. Silva *et al.* [35] made an instrument to survey the trustworthiness of IoT applications.

The standard used to arrange quality traits originates from ISO 25010, tied with some particular IoT properties gotten from the essential investigations key-wording. The building style of an IoT framework can have impact on quality characteristics;

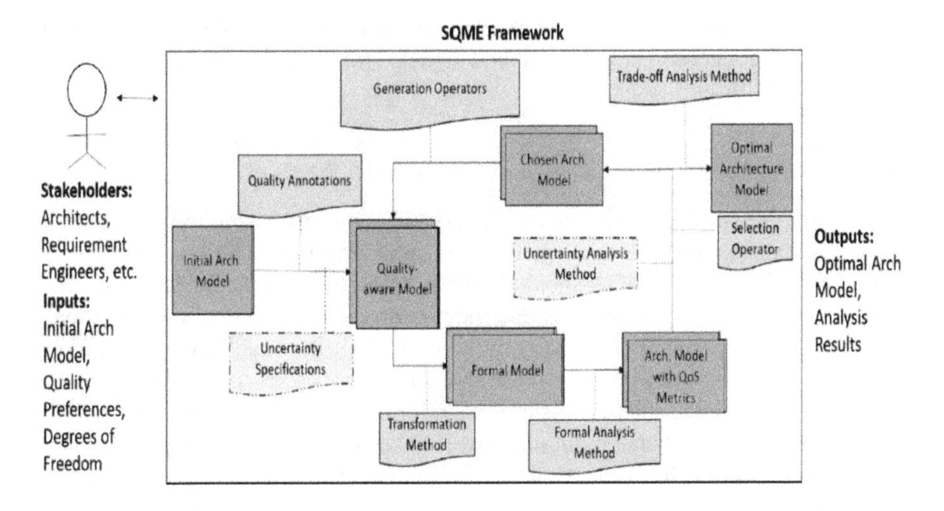

FIGURE 6.8 A basic overview of SQME framework [36].

however, it doesn't ensure every one of them. The most perceived quality ascribes that should be happy with an appropriate IoT design are adaptability, security, interoperability, and execution. Versatility is a basic trait, as IoT ought to be skilled to perform at a satisfactory level with this size of gadgets. Moreover, security increases a high worry in an IoT framework, where various parts and elements are associated with one another through a system. Interoperability helps heterogeneous segments of IoT to cooperate proficiently. Protection, accessibility, versatility, dependability, strength, and evolvability are situated in the lower level of concern. Versatility that is successful dealing with the disappointments and is a basic viewpoint, does not tend to unfathomably through essential investigations; however, it has a tremendous ability to be concentrated in future inquiries. Sedaghatbaf and Azgomi have presented a(Software Quality Modeling and Evaluation) SQME framework that evaluates the SA models automatically [36]. The SQME framework is shown in Figure 6.8.

6.3 THEORETICAL FOUNDATION FOR SOFTWARE INTEGRATED FRAMEWORK DESIGN FOR IoT-BASED APPLICATIONS

The various challenges and approaches of the related problem domains are discussed below. The HC approach is used for modeling LSS. The basic concept of the proposed approach is explained through an example. The conceptual model of SOA is discussed through the HC approach. The proposed method is illustrated through clinical decision support system (CDSS) as a case study.

There are various research challenges for modeling LSS. The research questions are: (i) What are the emerging research areas of *SOA*? (ii) What are the minimum number of parameters required to represent service composition? (iii) Which approach is suitable for designing LSS? (iv) What are the benefits of presenting LSS through the HC approach? (v) How are service dynamics and conceptual model adapted through the HC approach? (vi) What are the limitations of SSN ontology? (vii) How is the

SSN ontology adopted through hypergraph concept? and (viii) What properties of software services need to be presented and modeled for the design of software service applications?

6.3.1 Information Architecture for IoT-Based Applications

Nowadays, the number of service consumers (SCs) are increasing tremendously in LSS. Thus, different challenges are there for modeling LSS. The first challenge is that a suitable model for SOA should be both business-understandable and executable. Thus, the second challenge is that the precise representation of the large-scale service-oriented system (SOS) is very limited [15]. Thirdly, establishing connections and checking consistency between those characteristics is also a prime requisite [37]. In this context, the HC approach is presented for designing LSS. This approach has some benefits, such as being more scalable; the execution time of the program is less and is able to present the elements of LSS through the HC approach over the existing approaches [38].

However, several elements of IoT are not included in SSN ontology [39]. This ontology does not expose sensors to services that may characterize some new ideas in the IoT domain. Henson *et al.* [40] have presented semantically enabled sensor observation service (SemSOS) with high-level knowledge and low-level sensor readings. An ontological structure has been introduced for the presentation of connected objects in Web of Things [41]. The basic terminology of HC approach is defined as below.

HC is an assignment of colors (positive integer) to the vertices of the hypergraph so that the vertices of every hyperedge contain each color from the color set. The minimum number of colors is required to color the vertices of the hypergraph, known as the chromatic number ($\chi(H)$) of a hypergraph [42].

Figure 6.9 presents the modeling of LSS through the HC method. Here, $\chi(H) = 3$. $P = \{P_1, P_2, P_3\}$ be the set of service providers (SPs). The relation between SPs and services can be expressed as, $<P_1 \rightarrow S_1 \Lambda S_2 \Lambda S_3 \Lambda S_6>$, $<P_2 \rightarrow S_4 \Lambda S_5 \Lambda S_7 \Lambda S_{11}>$, and $<P_3 \rightarrow S_8 \Lambda S_9 \Lambda S_{10}>$. The mapping of LSS through HC approach is shown in Table 6.1.

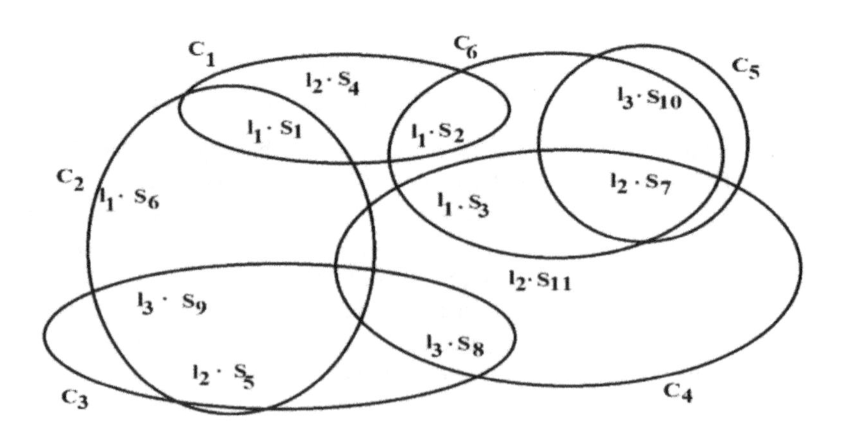

FIGURE 6.9 Modeling of SC-SP through a HC approach [38].

TABLE 6.1
Modeling of SC-SP through HC Approach [38]

Elements of SC-SP	Elements of HC
Service	Vertex
SC	Hyperedge
SP	Color
Number of SPs	Chromatic Number ($\chi(H)$)
Data value associated with the service for operation	Weight (W(H))
Maximum number of SCs consuming a service	Degree (Δ(H))

In the next table, the conceptual model such as information model (IM), action model (AM), behavioral model (BM), and artefacts model (AFEM) of SOS is mapped in the form of HC, as shown in Table 6.2 [38].

6.3.2 Evaluation of the Proposed Model Through a Case Study

The proposed approach is illustrated through a case study, i.e., CDSS [14]. CDSS consists of two services: Patient_Information_System and Consultation_Service. In this system, medicines act as input for the patient and the curable patients act as output of the system. The mapping from a CDSS to HC is shown in Table 6.3.

TABLE 6.2
Representation of Various Models of SOS Into HC [38]

Elements of Different Models	Elements of HC
IM	
Service_ID	Vertex number
SP	Color
Data	Weight
SC	Hyperedge
Choreography system	$W = 1$.
Orchestration system	$W = 0$.
AM	
Action type	Simple action ($V = 0$), composite action ($V = 1$)
Ordering number	Scheduling order of the services (Y)
Service	Vertex
Task	Vertex partition set
BM	
Effect	Result or output (O)
AFEM	
Constraint	C (Parameter)
Ordering number	Scheduling order of the services (Y)
Flag	Variable (F)

TABLE 6.3
Mapping from CDSS to HC [38]

Elements of CDSS	Elements of HC
Medicine	Vertex
Patient	Hyperedge
Medicine representative	Color
Minimum number of medicine representatives	Chromatic number ($\chi(H)$)
Maximum number of patients consume a single medicines	Degree of hypergraph ($\Delta(H)$)
Largest sum of medicines amount consumed by a patient	Weight of the hypergraph ($W(H)$)

It is a major challenge to implement the service composition in real life. The proposed approach will be extended for the cyber-physical system by including complex service dynamics [38]. There should be a standard language for portraying the IoT benefits in a proper manner. The objective of the service modeling approach is utilized to componentize the business component, since it ties services to the business goals. BPM ought to coordinate with IoT to achieve a dynamic context of a business process. There should be a system to associate the IoT foundation with BPM by incorporating IoT information to upgrade the dynamic of the business process in a methodological manner. Thus, a systematic method is required to propose this sort of framework [17]. However, it is not worthy to release the architecture of IoT without assessing the quality. In the next section, several research proposals are discussed in detail.

6.4 FUTURE RESEARCH DIRECTIONS

There are a couple of challenges in the business process requirements for modeling SOS frameworks. There are gaps existing between the representation of end-users' business goals and corresponding business scenarios. There merely exists any methodology for thorough investigation of business process requirements in the background of software-oriented requirement engineering [43].

6.4.1 Enrichment of Hypergraph-Based Approach for IoT-Based SA

Service performance needs in IoT are hard to fulfil because of the constraint of assets, as well as bandwidth, mode preparing capacities, and server limits. Congestion control should be appropriately modeled and analyzed. The idea of SSN ontology ought to be adopted with the elements of HC approach. This approach is utilized for demonstrating of large-scale SOA. The components of the SSN ontology, for example, device, event, observation, parameter, platform, and property, should be adopted with the components of services such as registry, service interface, service status, etc. [44]. This idea should be implemented practically and measured in terms of time and space. There are some minimum number of parameters required for an IoT framework to achieve integration and interoperability [45]. These parameters are scalability, ease of testing, ease of development, lightweight implementation, and service coordination [13].

6.4.2 Event-Driven Work Flow-Based Requirements Analysis for IoT-Based Applications

Business management can benefit from IoT concepts or devices. However, there are various challenges for the integration of IoT with business management. At first, there merely exists a suitable framework to integrate IoT infrastructure with BPM [18]. Secondly, the context layer and decision layer are still missing to obtain the added value of IoT awareness in BPM [18]. Thirdly, there is an absence of understanding of how information is utilized. Usage control has been presented in the 2000s [45]. Finally, it is a major challenge for enterprises to guarantee handling of critical data and also the information leak of private data.

The research questions are: (i) What are the standard ways to connect IoT infrastructure with business process? (ii) How to connect the IoT infrastructure with business process for obtaining the context-awareness? (iii) Who is responsible for leakage of private data with additional information in the IoT application? (iv) How data are used for processing in an enterprise?

6.4.3 Verification and Validation of the Service Composition Pattern

The developed approach [46, 47] validates the composition pattern through a high-level Petri-net-based approach using different tools. The author [48, 49] has described the verification plan for web services composition using enhanced stacked automata model. The verification and validation can be done by comparing the simulation result of the model. Less attention is paid towards the authentication of the composition pattern using high-level Petri-net.

6.4.4 Quality Evaluation Framework for IoT-Based SA

There may be a major challenge for designing software-intensive by considering different quality attributes, such as *performance*, *reliability*, and *security*. However, there are major difficulties in SA quality assessment by the existing methods [50]. At first, software quality properties are not autonomous from one another. Specifically, improving the architecture regarding one characteristic may influence another one. Secondly, there are no methodologies to suggest the improvement of architecture if the assessment results were not satisfactory. This task is usually done manually. Thus, it consumes more time and is error-prone [51]. Thirdly, there may exist a certain level of uncertainty in the architecture design stage, because of the absence of adequate information about the software system [52]. The exact value of certain parameters might be obscure right now. Fourthly, it is a major challenge to guarantee the quality of IoT framework rather than the quality of software. To guarantee the quality of the IoT framework requires a structure to assess every part individually and verify the expected output.

The research questions are: (i) How the quality aspect of IoT is handled in service perspective? (ii) How to provide the feedback of the evaluation results to the architectural model? (iii) What are the challenges in SA evaluation that are not addressed yet? (iv) How to generate the feedback of evaluation results to the architectural model automatically? A suitable framework is required for evaluating SA by considering different quality attributes.

6.5 CONCLUSION

A software architectural style is a labeled collection of architectural design decisions that are applicable to a particular context, constrain development within that context, and are yielding useful characteristics. These styles can be utilized to design the SA of a particular framework. SAs are unique to the system they represent. The components of large-scale SOA are mapped to the different components of the hypergraph. This methodology aids in finding the *number of services*, *service composition pattern*, and *scheduling order of the services* through parameters. This chapter has investigated the issues in incorporating IoT devices and frameworks with SSN ontology.

This chapter has highlighted the fundamental features for integration of SSN ontology with service. Future work incorporates the design of a framework that decreases the gap between IoT foundation and context-aware BPM. There should be an automation of quality evaluation process which reduces human effort, and that may reduce development costs.

REFERENCES

1. R. Capilla, A. Jansen, A. Tang, P. Avgeriou, and M. A. Babar, *10 Years of Software Architecture Knowledge Management: Practice and Future.* Journal of Systems and Software, vol. 116, pp. 191–205, 2016. DOI: 10.1016/j.jss.2015.08.054
2. J. Wan, S. Tang, Z. Shu, D. Li, S. Wang, M. Imran, and A. V. Vasilakos, *Software-Defined Industrial Internet of Things in the Context of Industry 4.0.* IEEE Sensors Journal, vol. 16, pp. 7373–7380, 2016.
3. P. Fremantle, *A Reference Architecture for the Internet of Things*; White Paper, WSO2: Mountain View, CA, 2014.
4. E. Cavalcante, M.P. Alves, T. Batista, F.C. Delicato, and P.F. Pires, *An analysis of reference architectures for the internet of things.* In Proceedings of the 1st International Workshop on Exploring Component-based Techniques for Constructing Reference Architectures, Canada, 2015.
5. M. Weyrich and C. Ebert, *Reference Architectures for the Internet of Things.* IEEE Software, vol. 33, pp. 112–116, 2016.
6. G. Campeanu, *A mapping study on microservice architectures of internet of things and cloud computing solutions.* In Proceedings of the 7th Mediterranean Conference on Embedded Computing (*MECO*), Montenegro, 2018.
7. H. Muccini and M.T. Moghaddam, *IoT architectural styles.* In European Conference on Software Architecture, January 2018. DOI: 10.1007/978-3-030-00761-4_5
8. J. Kiljander, A. D'elia, F. Morandi, P. Hyttinen, J. Takalo-Mattila, A. Ylisaukko-Oja, J.P. Soininen, and T. Cinotti, *Semantic Interoperability Architecture for Pervasive Computing and Internet of Things.* IEEE, vol. 2, pp. 856–873, 2014.
9. J. Taylor, A. Fish, J. Vanthienen, and P. Vincent, Emerging standards in decision modeling, an introduction to decision model notation, In *iBPMS: Intelligent BPM Systems: Impact and Opportunity BPM and Workflow Handbook Series.* Future Strategies, Inc., pp. 133–146, 2013.

10. F. Yan, S. Liu, and Z. Yao, Research on business process modeling and business process service based on Web service. In 2nd Symposium on Web Society, vol. 1, pp. 254–257, 2010.
11. P. Liu and B. Zhou, *Research on Workflow Patterns Based on jBPM and jPDL*, IEEE Pacific-Asia Workshop on Computational Intelligence and Industrial Application, vol. 2, pp. 838–843, 2008.
12. D. C. Petriu, M. Alhaj, and R. Tawhid, *Software performance modeling.* In Proceedings of the 12th International Conference on Formal Methods for the Design of Computer, Communication, and Software Systems: Formal Methods for Model-Driven Engineering, pp. 219–262, 2012.
13. O. Uviase and G. Kotonya. *IoT architectural framework: connection and integration framework for IoT systems.* 1st Workshop on Architectures, Languages and Paradigms for IoT, pp. 1–7, 2018.
14. A. K. Mandal and A. Sarkar, *Service oriented system design: Domain specific model-based approach*, 3rd International Conference on Computer and Information Sciences (*ICCOINS*), Kuala Lumpur, Malaysia, pp. 489–494, 2016.
15. A. Chaudhuri, S. Banerjee, and A. Sarkar, *Behavioral analysis of service oriented systems using event-B.* In 17th International Conference on Computer Information Systems and Industrial Management Applications (CISIM), Olomouc, Czech Republic, pp. 117–129, 2018.
16. M. N. Huhns and M. P. Singh, *Service-Oriented Computing: Key Concepts and Principles.* IEEE Internet Computing, vol. 9, no. 1, pp. 75–81, 2005.
17. Rongjia Song, Jan Vanthienen, Weiping Cui, Ying Wang, and Lei Huang, *Context-aware BPM using IoT-integrated context ontologies and IoT-enhanced decision models*, 21st IEEE Conference on Business Informatics, Moscow, Russia, pp. 541–550, 2019.
18. S. Dustdar and W. Schreiner, *A Survey on Web Services Composition*, International Journal on Web and Grid Services, vol. 1, no. 1, pp. 1–30, 2005.
19. M. Compton, P. Barnaghi, L. Bermudez, R. Garcia-Castro, O. Corcho, S. Cox, J. Graybeal, M. Hauswirth, C. Henson, A. Herzog, V. Huang, K. Janowicz, W. David Kelsey, D. Le Phuoc, L. Lefort, M. Leggieri, H. Neuhaus, A. Nikolov, K. Page, A. Passant, A. Sheth, and K. Taylor. *The SSN Ontology of the W3C Semantic Sensor Network Incubator Group.* Web Semantics: Science, Services and Agents on the World Wide Web, vol. 17, pp. 25–32, 2012.
20. K. Janowicz, A. Haller, S. J.D. Cox, D. Le Phuoc, and M. Lefrancois, *SOSA: A Lightweight Ontology for Sensors, Observations, Samples and Actuators.* Web Semantics: Science, Services and Agents on the World Wide Web, vol. 56, pp. 1–10, 2018.
21. H. Neuhaus, and M. Compton. *The semantic sensor network ontology: a generic language to describe sensor assets.* In AGILE Workshop Challenges in Geospatial Data Harmonisation, Hannover, Germany, 2009.
22. Denis Kleyko, Pattern recognition with vector symbolic architectures, Ph.D. dissertation, Dept. of Computer Science and Electrical Engineering, Lulea University of Technology, Lulea, Sweden, 2016.
23. Node-RED: Flow-based programming for the Internet of Things, JS Foundation, https://nodered.org/, 2013.
24. R. Tomsett, G. Bent, C. Simpkin, I. Taylor, D. Harborne, A. Preece, and Raghu K. Ganti. *Demonstration of dynamic distributed orchestration of node-RED IoT workflows using a vector symbolic architecture.* In 5th IEEE International Conference on Smart Computing (SMARTCOMP), pp. 464–467, 2019.
25. Julian Schutte, and Gerd Stefan Brost, *LUCON: data flow control for message-based IoT systems.* In 17th IEEE International Conference on Trust, Security and Privacy in Computing and Communications/12th IEEE International Conference on Big Data Science and Engineering (*TrustCom/BigDataSE*), pp. 289–299, 2018.

26. T. F. J. Pasquier, J. Singh, D. M. Eyers, and J. Bacon. *CamFlow: Managed Data-Sharing for Cloud Services.* IEEE Transactions on Cloud Computing, vol. 5, no. 3, pp. 472–484, 2015.
27. K. Dar, A. Taherkordi, H. Baraki, F. Eliassen, and K. Geihs, *A Resource Oriented Integration Architecture for the Internet of Things: A Business Process Perspective.* Pervasive and Mobile Computing, vol. 20, no. 7, pp. 145–159, 2015.
28. S. Meyer, A. Ruppen, and C. Magerkurth, *Internet of things-aware process modeling: integrating IoT devices as business process resources.* In Proceedings of the 25th International Conference on Advanced Information Systems Engineering (CAiSE), Valencia, Spain, 2013.
29. H.-H. Chiu, and M.-S. Wang, *A Study of IoT-Aware Business Process Modeling.* International Journal of Modeling and Optimization, vol. 3, no. 3, pp. 238–244, 2013.
30. Mi Kim, *A Quality Model for Evaluating IoT Applications.* International Journal of Computer and Electrical Engineering, vol. 8, no. 1, pp. 66–76, 2016.
31. J. H. Liu, W. Q. Tong, *Adaptive service framework based on grey decision-making in the Internet of Things.* In Proceedings 6th International Conference on Wireless Communications, Networking and Mobile Computing (*WiCOM*), Chengdu, China, 2010.
32. X. Zheng, P. Martin, K. Brohman, and L. D. Xu, *Cloudqual: A Quality Model for Cloud Services.* IEEE Transactions on Industrial Informatics, vol. 10, no. 2, pp. 1527–1536, 2014.
33. M. Ahmad, *Reliability models for the internet of things: a paradigm shift.* In Proceedings IEEE International Symposium on Software Reliability Engineering Workshops, pp. 52–59, 2014.
34. A. Karkouch, H. Mousannif, H. A. Moatassime, and T. Noel, *A model-driven architecture-based data quality management framework for the internet of things.* In Proceedings 2nd International Conference on Cloud Computing Technologies and Applications (*CloudTech*), pp. 252–259, 2016.
35. I. Silva, R. Leandro, D. Macedo, and L. A. Guedes, *A Dependability Evaluation Tool for the Internet of Things.* Computers and Electrical Engineering, vol. 39, no. 7, pp. 2005–2018, 2013.
36. A. Sedaghatbaf, and M. A. Azgomi, *SQME: A Framework for Modeling and Evaluation of Software Architecture Quality Attributes.* Software and Systems Modeling, vol. 18, pp. 2609–2632, 2019.
37. Shreya Banerjee, Shruti Bajpai, Anirban Sarkar, Takaaki Goto, and Narayan C Debnath, *Ontology driven meta-modelling of service oriented architecture.* In 3rd International Conference on Communication, Management and Information Technology, Warsaw, Poland, 2017.
38. S. K. Mishra, R. A. Haraty, N. C. Debnath, and A. Sarkar, *A hypergraph coloring based modelling of service oriented system.* In IEEE 17th International Conference on Industrial Informatics (INDIN), Helsinki, Finland, pp. 1349–1350, 2019.
39. Henry Muccini and Mahyar T. Moghaddam, *IoT architectural styles a systematic mapping study.* In 12th European Conference on Software Architecture (*ECSA*), Madrid, Spain, pp. 68–85, 2018.
40. C. A. Henson, Josh K. Pschorr, Amit P. Sheth, and K. Thirunarayan. *SemSOS: semantic sensor observation service.* In International Symposium on Collaborative Technologies and Systems, pp. 44–53. 2009.
41. Wei Wang, Suparna De, Gilbert Cassar, and Klaus Moessner. *Knowledge Representation in the Internet of Things: Semantic Modelling and its Applications.* Automatika, vol. 54, pp. 388–400. 2012.
42. N. Gyorgy and Cs. Imreh, *Online Hypergraph Coloring.* Information Processing Letters, vol. 109, no. 1, pp. 23–26, 2008.

43. Shreya Banerjee and Anirban Sarkar. *A Requirements Analysis Framework for Development of Service Oriented Systems.* ACM SIGSOFT Software Engineering Notes, vol. 42, pp. 1–12, 2017.
44. J. Huang, D. Du, Q. Duan, Y. Sun, Y. Yin, T. Zhou, and Y. Zhang, *Modeling and analysis on congestion control in the Internet of Things.* In 48th IEEE International Conference on Communications (*ICC*), Sydney, Australia. pp. 434–439, 2014.
45. J. Park and R. Sandhu. *The* $UCON_{ABC}$ *Usage Control Model.* ACM Transaction on Information System Security, vol. 7, no. 1, pp. 128–174, 2004.
46. W. L. Dong, H. Yu, and Y. B. Zhang, *Testing BPEL-based web service composition using high-level Petri-nets.* In 10th IEEE International Enterprise Distributed Object Computing Conference (EDOC), pp. 441–444, 2006.
47. D. Nagamouttou, I. Egambaram, M. Krishnan, and P. Narasingam, *A Verification Strategy for Web Services Composition Using Enhanced Stacked Automata Model.* Springer Plus, vol. 4, no. 1, 2015.
48. C. S. Chen, C. H. Lin, and H. Y. Tsai. *A Rule-Based Expert System with Colored Petri-net Models for Distribution System Service Restoration.* IEEE Transactions on Power Systems, vol. 17, no. 4, 1073–1080, 2002.
49. Ali Sedaghatbaf and Mohammad Abdollahi Azgomi, *SQME: A Framework form Modeling and Evaluation of Software Architecture Quality Attributes.* Software and Systems Modelling, vol. 18, pp. 2609–2632, 2019.
50. A. Martens, H. Koziolek, S. Becker, and R. Reussner, *Automatically improve software architecture models for performance, reliability, and cost using evolutionary algorithms.* In Proceedings of the First Joint WOSP/SIPEW International Conference on Performance Engineering, pp. 105–116, 2010.
51. C. Trubiani, I. Meedeniya, V. Cortellessa, A. Aleti, and L. Grunske, *Model-based performance analysis of software architectures under uncertainty.* In Proceedings of the 9th International ACM Sigsoft Conference on Quality of Software Architectures, pp. 69–78, 2013.
52. M. Autili, V. Cortellessa, D. Di Ruscio, P. Inverardi, P. Pelliccione, and M. Tivoli, *EAGLE: engineering software in the ubiquitous globe by leveraging uncertainty.* In Proceedings of the 19th ACM SIGSOFT Symposium and the 13th European Conference on Foundations of Software Engineering, pp. 488–491, 2011.

7 Security Issues and Challenges in IoT

Sandeep Mahato, Kailash Chandra Mishra, and Subrata Dutta
NIT Jamshedpur, Jharkhand, India

Sujoy Mistry
Maulana Abul Kalam Azad University of Technology, West Bengal, India

CONTENTS

7.1 Introduction 117
7.2 The Open Issues and Challenges 117
7.2.1 IoT Issues as per Different Perspective 117
7.2.2 Challenges and Open IoT Issues in Domain of Privacy 118
7.2.3 Challenges and Open IoT Issues in Secure Routing and Forwarding 119
7.2.3.1 Significant Problems in Secure Routing and Forwarding 119
7.2.4 Open Issues in Secure Routing and Forwarding 119
7.2.5 Key Challenges to Maintaining Robustness and Resilience 120
7.2.6 Key Challenges in Denial of Service (DoS) and Insider Attacks 120
7.3 IoT Security Threats 121
7.3.1 Denial of Service (DoS) 121
7.3.2 Physical Harm 121
7.3.3 Eavesdropping 121
7.3.4 Node Capture 121
7.3.5 Controlling 121
7.4 IoT Attack Vectors and Security Vulnerabilities 122
7.4.1 Memory for IoT Device 122
7.4.2 Web Integration for IoT 122
7.4.3 Services for IoT Devices Network 122
7.4.4 IoT Device Cloud Connectivity 122
7.4.5 IoT Device Software Update 122
7.4.6 IoT Device to Device Connectivity 123
7.4.7 IoT AAA Services 123
7.4.8 IoT Data Storage Methods 123

7.5 Potential Attacks and Countermeasures in IoT or Smart Objects ... 123
7.5.1 Cloud Domain Security Attacks ... 124
7.5.1.1 Hidden-Channel Attacks ... 124
7.5.1.2 VM Migration Attacks ... 124
7.5.1.3 Theft of Service Attack ... 125
7.5.1.4 VM Escape Attack ... 125
7.5.2 Fog Domain Security Issues and Attacks ... 126
7.5.2.1 Authentication and Trust Issues ... 126
7.5.2.2 Higher Migration Security Risks ... 126
7.5.2.3 Higher Vulnerability to DoS Attack ... 126
7.5.2.4 Further Risks to Health Due to Use of Containers ... 127
7.5.2.5 Privacy Issues ... 127
7.5.3 Sensing Domain Security Attacks ... 127
7.5.3.1 Jamming Attack ... 127
7.5.3.2 Vampire Attack ... 127
7.5.3.3 Denial of Sleep ... 128
7.5.3.4 Flooding Attack ... 128
7.5.3.5 Carousel Attack ... 128
7.5.3.6 Stretch Attack ... 128
7.5.3.7 Selective-Forwarding Attack ... 129
7.6 Security Requirements ... 130
7.6.1 Confidentiality ... 130
7.6.2 Integrity ... 130
7.6.3 Authentication ... 130
7.6.4 Availability ... 130
7.6.5 Authorization ... 130
7.6.6 Freshness ... 130
7.6.7 Non-Repudiation ... 130
7.6.8 Forward Secrecy ... 130
7.6.9 Backward Secrecy ... 131
7.7 IoT Security Challenges ... 131
7.7.1 Protocol and Network Security ... 131
7.7.2 Identity Management ... 131
7.7.3 Privacy ... 131
7.7.4 Trust and Governance ... 132
7.7.5 Fault Tolerance ... 132
7.8 Privacy and Trust ... 132
7.9 IoT Security Mechanism ... 133
7.9.1 Authentication ... 133
7.9.2 Encryption ... 134
7.9.3 Trust Management ... 134
7.9.4 Secure Routing ... 134
7.9.5 New Technologies ... 134
References ... 135

7.1 INTRODUCTION

Internet of Things (IoT) is a term that describes a network of sensors, objects, and intelligent nodes and technologies, which can communicate with each other and do business with their neighbors to achieve some common goals without human intervention. IoT is based on wireless sensor networks (WSNs), where sensor nodes are connected and communicate with each other with minimal or no human interactions. In certain cases, such entities are heterogeneous, such as individuals, machines, books, cars, and other devices that could be installed, addressed, and read on the internet. IoT can be summed up as "a global network of linked entities." They are able to connect with all of the other network organizations, and receive and provide services at anytime, anywhere [1].

The entities in IoT can be remotely controlled for the desired functionality. The sharing of information between devices takes place via the network with standard communication protocols. The intelligent connected bodies or "things" range from simple, small, and wearable accessories to large machines with sensor chips.

According to [2], the IoTs are like real-time systems, which collect data through sensors and interact in the network to give users the ability to share, find, and take action. IoTs have identified their application area as a home automation system as a smart home where they can control household devices with their laptops and mobile phones, intelligent transportation system for traffic monitoring and reporting traffic jams, traffic violations and accidents, weather forecasting and natural disaster prediction with continuous monitoring of temperature change, toxic gases and material contamination for pollution, healthcare by monitoring activities of patients, medicine intake, and their various health parameters, and surveillance of objects, animals, and people with alarming systems. Many more applications and facilities are possible [3]. In addition to the rapid growth of IoT applications and devices, major challenges, such as security and privacy issues with vulnerabilities and other threats, have also improved more than ever. Several manufacturers of IoT products, such as cloud providers and researchers, are working on the design of safety systems and emerging vulnerabilities. While IT security and privacy continue to be tackled in research, most studies are not applicable and are only in their early stages. There are still many issues available.

7.2 THE OPEN ISSUES AND CHALLENGES

Despite the immense promise of IoT in a range of fields and applications, researchers now face a number of obstacles and concerns related to IoT. If we consider data, the collection of data from sensors and other IoT system appliances may be large enough to handle. Sometimes the data needs to be processed in real time in IoT, and then the sensor takes its own decision without sufficient storage to keep all the data collected. Many researchers addressed even more open topics and challenges.

7.2.1 IoT Issues as per Different Perspective

In accordance with C. W. Tsai *et al.* [3, 4], IoT issues can be viewed as perspectives on infrastructure, data management, and computational intelligence. IoT, with characteristics such as heterogeneity and decentralization, encompasses different standardizations, communication between interfaces, the provision of a number of IP addresses, and the management of decentralization issues under the infrastructure perspective. The collection of data from appliances and extraction of useful information from collected data when the number of appliances increases, along with common issues of data presentation between devices and appliances, are issues of data management perspective. Once IoT is sufficiently efficient in data management, another challenge is to make smart appliances and make useful decisions. Computational intelligence technologies such as data mining are still at an early stage of growth. Advanced computational technologies are expected to be used for better IoT services. Some future IoT issues have also been discussed as cloud computing and social network perspectives. When cloud computing is part of the internet, the integration of cloud and IoT and the transmission of data between these two different environments leads to considerable problems. Improving IoT performance through the use of digital social behavior is a challenge from the perspective of the social network.

7.2.2 Challenges and Open IoT Issues in Domain of Privacy

As per S. Hameed *et al.* [3, 5], preserving privacy in IoT is a major security concern. Profiling and monitoring, localization, monitoring, and safe data transmission are some of the challenges to maintain privacy in IoT. An identification linked to another individual is a hazard in profiling and monitoring. A major challenge still lies in using some preventive measures to avoid such activities. In localization and monitoring, a program attempts to keep track of an entity's position across space and time. The task is to design, describe, and enforce IoT interaction protocols. To avoid unauthorized access to information about people and objects in secure data transmission, a protected transmission channel is necessary to ensure that data is transmitted in a safe manner. There are several open problems in the IoT privacy domain as per author, such as a comprehensive system for privacy protection, which guarantees IoT privacy for different applications. A generic, lightweight cryptographic privacy algorithm can be implemented as a solution. Context away privacy policies, as well as innovations, such as privacy policies that concentrate on users and remote environments, context-centered and self-adaptive policies, as well as intelligent data protection protocols, have been proposed. Data access management and data management systems are also needed. Privacy-preservation of game theory rewards where game theory was used to examine the privacy of the venue. A significant challenge for open research is how to use game theory to enable the IoT-architecture to safeguard its privacy protocols. Virtualization network and SDNsAdaptation for Network Virtualization for privacy security has evolved as a possible approach to a large number of data processed in IoT and cloud management. The Network Defined Software (SDN) was recently introduced as a virtualization model for the network. SDN also controls the network by centralizing Routing and Transmission functions at a key point known as the controller. According to [6], the network operator and administrator will thus enforce confidentiality in the network with ease of network management by enabling SDN in some advance programmable networks.

7.2.3 Challenges and Open IoT Issues in Secure Routing and Forwarding

7.2.3.1 Significant Problems in Secure Routing and Forwarding

Secure route establishment: One of the most important problems is to securely create the routing protocol to transmit data in IoT. This protocol would set up a secure route and safe route between the nodes. Based on routing data, the calculations should be lightweight and serve the IoT networks with low power capacity. K. T. Nguyen *et al.* [7] gave a new classification of existing protocols, based on their main method of setting up a secure communication channel.

Malicious nodes isolation: Another task is to detect nodes and design strategies isolated from IoT networks easily and robustly. In order to reduce or remove inconsistencies in the routing process, the solution protocol should be able to insulate malicious and misbehaving network nodes. Current IoT routing protocols are unsafe, as most IoT networks organize themselves and do not interact with people. Malicious nodes can therefore be added fairly quickly to the IoT network, meaning that an IoT network requires a protocol to identify malicious nodes in the network, using techniques and mechanisms, until malicious activities begin or stop them from being linked. The earlier techniques of detection primarily identify system anomalies based on system logs or events, while the latter focus on network threat detection through traffic assessment of filtering and sampling [8].

Security protocol self-stabilization: The self-stabilization protocol means that it is naturally possible to recover in a period without human interaction from unintended events and problems.

Preservation of location privacy: The privacy of location of IoT devices in the IoT network should be preserved. It should thus be able to preserve the privacy of location for a secure routing protocol. The new Dummy Location privacy algorithm (DLP) is introduced by Sun Gang *et al.* [9] and takes into account the balance of cost (i.e., time complicated) and the privacy requirements of the user.

7.2.4 Open Issues in Secure Routing and Forwarding

Routing protocol design of iot network with focus on performance: Although numerous research has addressed the issues of secure routing and forwarding, the efficiency of the IoT network has not been found for such research. The identification of attacks on the network is complex, for example with IDS. The resource limitation of IoT devices is not considered in these frameworks. Lightweight IDS, however, will help us detect malpractice in the IoT network and attempt to mitigate routing attacks within the IoT network. Lightweight IDS is proposed by C. Ioannou and V. Vassiliou [10] to detect the presence of an attack locally at every restricted node, based on the Binary Logistic Regression (BLR). Network researchers are expected to develop a new lightweight IDS architecture.

Efficient and comprehensive routing operation management: Controlling routing operations efficiently and thoroughly. Besides lightweight IDS for IoT, the IoT network can be centrally managed and helpful in the overall network monitoring. Researchers do need high-level IoT network grassroots security and fast-response routing control policies for security threats. A mechanism such as SDN, which centralizes the controller and lets researchers securely route data through the IoT network, is appropriate. As per [11] SDN may play an important role with the blocks or rate of limiting suspicious flows in dynamical flow control and attack mitigation. Advanced security solutions are therefore essential with SDN, allowing IoT network data to be transmitted while protecting its integrity.

7.2.5 Key Challenges to Maintaining Robustness and Resilience

Attack tolerance: IoT networks require sophisticated, stable network architectures that are basically tolerant of malicious attacks and other intrusions.

Early attack detection: The IoT network must provide early attack detection features and protocols to prevent damage and contamination across the network.

Quick failure recovery: Time failure recovery in the IoT network is important. Long-term IoT services disorders can be life-threatening, especially for healthcare services and applications in disaster management. The resource management middleware for IoT network must therefore identify faults and fix the situation in time. Middleware can use event-based, service-focused, tuple-space, VM-based, database-oriented, and application-specific approaches, according to their design approaches [12]. To solve IoT system problems, a number of other potential solutions remain. One potential solution calls for a duplication of resources that make it costly.

7.2.6 Key Challenges in Denial of Service (DoS) and Insider Attacks

Effective resource identification of attack: DoS attacks are difficult to recognize, therefore efficient DoS detection mechanisms are needed before the attack is started. Various proposals are available to deal with DoS. IoT DDoS detection is a tough task because IoT network and traffic features differ considerably from the conventional network. In terms of power, storing, anddata collection and processing, resource-efficient DDoS detection and mitigation techniques, there are some limits to IoT devices. For example, [13] exploiting vulnerability in DoS or DDoS and hiding true IP addresses affects IoT creation integrity. These strategies should be standardized so that traffic monitoring can take place centrally within the IoT network. In the event of DDoS attacks, certain probabilistic approaches are useful. Such methods, on the other hand, can be spread when the probability of a DDoS attack on IoT network is deduced cooperatively by many IoT tools.

Resource efficient countermeasures: Effective countermeasures of resources. To mitigate this attack, countermeasures are required if a DoS attack is detected. The resource-restricting of IoT networks requires lightweight and energy-efficient counter measurement strategies.

Efficient resource detection for insiders: Only authorized IoT nodes should be part of the IoT network to protect them from insider attacks. Malicious insider detection techniques within the IoT network must be effective and must react in real time. These malicious persons may also leak sensitive data across a network, compromising nodes or distorting the functioning of an IoT network through attacks like a DDoS attack. Another highly destructive scenario can arise.

7.3 IoT SECURITY THREATS

As per R. R. Roman *et al.* [14] in IoTs, an attacker may control part of the system, but it is not possible for an assailant to completely control the entire system due to the innately distributed existence of IoTs. It helps an intruder to be both "within" and "external." Therefore, these assailant models are marked as safety risks.

7.3.1 Denial of Service (DoS)

The IoT can be released with a large number of DoS attacks. However, in addition to traditional internet DOS attacks, e.g., chain jamming, the existing wireless communication system can be attacked on most data acquisition networks. Malicious domestic aggressors who take possession of certain infrastructures may cause even more confusion.

7.3.2 Physical Harm

This hazard can be regarded as part of the DoS hazard. Active attackers can lack technical expertise in this form of threat and can only interfere with provision of IoT services by damaging the actual "stuff." This is a practical IoT assault, because objects can easily be reached by anyone (e.g., streetlights). The attacker simply aims to damage the hardware module if this is not possible.

7.3.3 Eavesdropping

In this model of attacker, passive attackers target and block different communication channels to eliminate information (e.g., mobiles, wired networks, internet) from the flow of information. In this network, the information circulating can be derived from an internal network intruder, which gains access to the resources within.

7.3.4 Node Capture

As stated earlier, materials (e.g., street lights, appliances) are in a certain setting. A successful attacker may physically attempt to retrieve the data without it being destroyed. Alternatively, active attackers can target other devices and components like data processing and data storing agencies that store data.

7.3.5 Controlling

As long as there is a path to attack, active assailants can try to partly or completely manipulate an IOT entity. The degree to which such attackers can cause harm depends in particular on: (a) importance of data handled by the particular item and system, and (b) the services rendered by the item and appliance. The extent of damage caused by such assailants mainly depends on (a) the value of the data handled by the element or system, and (b) the services supplied by the element and system concerned. As per K. M. Sadique *et al.* [15], IoT security risks have been identified as (1) distrust of IoT devices by distrusting manufacturers, (2) replacement of things with malicious inferior quality things, (3) middle attack by man due to lack of proper authentication and authorization mechanisms during transmission, (4) malicious code replacement by an attacker; M. Sadique *et al.* [15] and Internet Engineering Task Force (IETF) IoT safety threats.

7.4 IoT ATTACK VECTORS AND SECURITY VULNERABILITIES

According to J. Ahamed and A. V. Rajan [16], IoT security threats continue to grow and mature, and it is a challenge for developers and researchers to address these threats. It is therefore necessary to try all the different ways to protect IoT devices with internet-enabled services. To accomplish this task, the various attack vectors must be studied.

7.4.1 Memory for IoT Device

Many IoT systems have little room for storage, so the applications need external memory. With their open architecture, most IoT equipment would be publicly exposed. This would open up a number of vulnerabilities to the system, for example, by revealing the device ID or the IoT serial number. Compromising and system interference would also have to be addressed.

7.4.2 Web Integration for IoT

Most IoT systems can contain web integration to connect to database servers. Cross-site injection and scripting are among the key safety risks for IoT systems, some of which may have an impact on IoT's web interface.

7.4.3 Services for IoT Devices Network

Another IoT device combat vector is a network services IoT device. IoT devices are very difficult to execute high-level encryption algorithms, making the devices vulnerable to data attacks on discovery. Unable to identify problems with normal data identification, a device can lead to different types of DDoS attacks. Sensor nodes are not needed by resource constraints, like computing power and data storage capacity to perform load testing or integrity tests, which renders the system vulnerable.

7.4.4 IoT Device Cloud Connectivity

Many IoT devices have cloud infrastructure connections that have major concerns about safety. Attackers can analyze the data through compromise of the cloud architecture. If a cloud attacker is able to compromise, one malware could be dangerous because it can be loaded simultaneously into several IoT devices.

7.4.5 IoT Device Software Update

The updating of the program is one process that in an IoT device can never be disrupted or skipped. Manually upgrading the patches one by one would still not be feasible for each IoT system. A cloud-based approach will work. However, there are still security risks to the cloud, and upgrading IoT devices' software patches is still a research area.

7.4.6 IoT Device to Device Connectivity

Another vector of attack is device-to-IoT computer communication. Hackers use rogue devices to communicate and collect information that is confidential and sensitive with the existing IoT system. For the safeguarding of the IoT systems in terms of the device's link, efficient authentication and encryption schemes are required.

7.4.7 IoT AAA Services

Services such as Authentication, Authority, and Accounting (AAA) would be a difficult task in IoT. Due to the distributed nature of IoT systems, the AAA device would be more challenging. It still is not clear how the AAA services are performed on the system by a single device or device group. There must be a secure way to communicate the authentication keys and tokens without divulging. The authentication method shall take into account device by device, equipment by device, and cloud authentication tool.

7.4.8 IoT Data Storage Methods

An external system safety vector can be found in IoT devices. Because of restricted computing power and storage space, IoT system management data are unencoded, in order for IoT apps not to verify data integrity. This prevents secrecy and authorization of the device. Most IoT devices could only support symmetric encryption where you can use static keys to encrypt and decrypt. Once the key has been compromised, the system can also affect the current and future scenarios.

7.5 POTENTIAL ATTACKS AND COUNTERMEASURES IN IoT OR SMART OBJECTS

The key security attacks in three domains of IoT, such as Cloud domain, fog domain, and sensing domain, have been identified and defined by A. Rayes and S. Salam [17]. For this broad number of servers with fast data processing, Cloud domain includes the IoT applicator and IoT services. In the fog domain, the fog device performs data acquisition operations, including aggregation, preparation, and storage. Fog devices are also linked to another in order to manage communication between intelligent objects and to cooperate how this fog device is responsible for handling what objects will alter their position over time.

All smart objects and devices that can feel the surroundings and record sensed data on fog domain devices are part of the sensing domain. Here are discussed some of the possible security threats in IoT.

7.5.1 Cloud Domain Security Attacks

7.5.1.1 Hidden-Channel Attacks

In addition to a logical separation between the VMs running on the same server, certain hardware components like cache are shared between these VMs. It can give the VMs live on the same server a chance for data leakage. Various countermeasures to avoid hidden channel attacks may be adopted. Any of them will be discussed. Hard isolation, which is essential for maintaining a high level of isolation among VMs under this precautionary technique, will be discussed. For this, the hardware or software for each VM is available with a separate dedicated cache. Either way, Cache Flushing technology flushes the cache sharing when cache delivery is transferred from one VM to the other. The defect is that the VMs on the server often experience a performance breakdown if that cache is vacuumed every time a VM switch occurs, which increases time to access and retrieve information. It's also hard to know whether a cache or memory collected the data, using noisy access time technology, adding noisy data to the time necessary to retrieve the data.

This makes it impossible for a malicious VM in the case of another VM that shares a server to recognize the segments of the cache. This has, of course, a cost because the time to recover the data cost for the collected data is slightly delayed because of the noise (variable time delay). The restricting cache speed is a method for reducing the amount of data to be drawn through VMs, by restricting how much the cache is transferred from one VM to the next.

If the cache is not transferred from a VM to another one too early in this strategy, then the cache content will be changed drastically by the VM that has a cache. This makes it difficult for another VM to gain a thorough understanding of the data accessed by the previous VM in cache testing.

7.5.1.2 VM Migration Attacks

Virtualization allows Live VM relocation, which enables a VM to be transferred transparently from a server to another server. Some of these types of attacks are as follows.

7.5.1.2.1 Control Plane Attacks

This type of attack targets the module that can manage migration on a server called the hypervisor migration module. The hacker is able to steal the server with a software migration module bug and control the migration module completely. Such attacks are mentioned.

Overflowing for migration: The attacker transfers all VMs stored on the hacked server to a victim server with inadequate resources for hosting all moving VMs. This triggers a service denial of the VM applications, as all host VMs resulting in VM performance loss and VM collision do not comply with the specifications.

Advertising for fake assets: The compromised server says that it's a huge slack tool like other free services. This attracts other servers to release certain VMs to the hacked server, thus distributing the workload on the cloud servers. After moving VMs from other normal servers to the hacked server, it is easier for the attacker to break into discharged VMs with vulnerabilities. Such VMs are then installed on a server under the attacker's power.

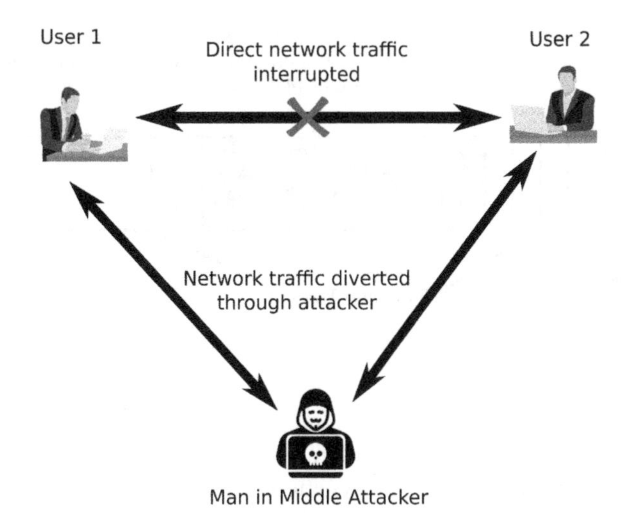

FIGURE 7.1 Man-in-the-middle attack.

7.5.1.2.2 Attacks on Data Plane

A second form of attacks on VM migration are triggered by this kind of attack, which threaten the network links over which the VM is transferred from one server to another. Any of the following are mentioned.

Sniffing attack: The intruder sniffs the initial and target packets and reads transmitted data files.

Man-in-the-middle attack: In a similar attack to one usually sent when a VM switches from one server to another, the attacker generates a free ARP response packet. This generated ARP packet informs the routing devices that the victim's VM address is changed to the attacker's physical address.

The incoming packets are then forwarded to the assailant's current physical address. Figure 7.1 demonstrates the middleman attack. If a protection function does not preserve the integrity of packets, an attacker can amend the contents of the received packages. As per [18], attackers secretly reveal and likely alter contact between two parties who feel that they communicate directly with one another.

7.5.1.3 Theft of Service Attack

In this attack, a deceptive VM acts to make the VM manager devote more resources to it than the amount it should receive. This increased resource allocation to the misty VM is at the detriment of other VMs using the same server that the malicious VM, where those victims of the VM get less assistance than they would actually get. Figure 7.2 shows service theft. In order to handle service theft, two countermeasures were proposed. The first remedy is to log the start and end times accurately when each VM uses precise clocks for the cores. Randomizing the sampling times is another countermeasure.

7.5.1.4 VM Escape Attack

Virtual machines are configured such that any VM on the same server runs and isolates it from other VMs, to prevent VMs accessing data from other VMs on the

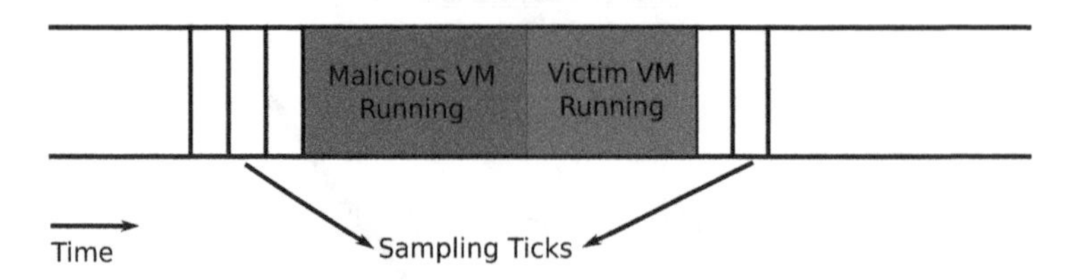

FIGURE 7.2 Theft of service attack.

same server. In reality, however, software bugs can break the isolation. The malicious VM may get root access to the entire system where a VM escapes the monitor layer and enters the hardware of the system. It offers full VM power over all VMs on the hacked server. An extra separation layer among the devices and the monitor can be added to counteract VM escape attack by nested virtualization, so that the unwelcome VM does not obtain root rights, although it bypasses a VM monitor layer. Certain options are software to avoid attacks from VM escape.

Insider attacks: For all attacks that have been previously discussed, the cloud data center managers were treated as trusted entities and were focused on attacks from the other unpleasant VMs hosted at the cloud data center. Nevertheless, such sensitive software may have grave concerns regarding the capacity of cloud data center administrators to access and change collected information [17].

7.5.2 Fog Domain Security Issues and Attacks

7.5.2.1 Authentication and Trust Issues

Fog systems are likely to be owned by many lesser-known companies. Authentication to identify the owner of the fog equipment when assigning an intelligent object to a fog equipment is critical for protection. It is also necessary with authentication to determine whether or not to trust the owner of the fog system. Confidence is a significant problem because the smart object can be distributed to multiple fog devices and can be used by different organizations as its location changes over time.

7.5.2.2 Higher Migration Security Risks

VM migration is common for both fog and cloud environments. The migrated VMs are transported across the internal network of the cloud data center. The fog device migrates to another computer through the internet. If a VM is transferred from a fog system to another, there is a greater likelihood of swapping VMs to reveal insecure network connections or network routers. It is therefore necessary to encrypt the migrated VM and to authentically exchange migration messages between fog devices.

7.5.2.3 Higher Vulnerability to DoS Attack

Less fog system machine ability makes it easier for attackers to immerse fog systems in denial of service (DoS) attacks. Compared to cloud data centers, where there is a large number of servers with a high processing power, attacks by DoS are no simpler.

7.5.2.4 Further Risks to Health Due to Use of Containers

The container in the fog solution is more computerized than VMs for the delegation of the resources specifications for each connected object to have a larger connection between various objects. The fog tool will serve a bigger number of items, if there is a small container overhead. If a container dedicated to objects from different users has the same operating system, severe safety issues will lead to the leakage of data and significantly hide the fog device being shared.

7.5.2.5 Privacy Issues

As already mentioned, any intelligent object must be linked to one of the nearby fog devices. This means that a fog device can track users or know their movement patterns. By taking care of the location of all connected intelligent objects, it can break the private space of users carrying other objects. Strategies have to be built to make it more difficult for fog resources to track the position of intelligent devices over time. The wireless signal processing revolution allowed the movement of people, objects and their position, and lips and animals to recognize and interpret their heartbeats through their wireless signals that communicate between sensors and fog domains. The revolution in wireless signal processing has facilitated the identification.

7.5.3 Sensing Domain Security Attacks

The sensing domain can be targeted many times. There are some of them listed below.

7.5.3.1 Jamming Attack

This type of attack leads to an interruption of service and has two types: jamming of the receiver when the assault is targeting the OSI stack physical domain of the receiver, where the receiver produces a signal (called the jamming signal) when a deceptive user (called the jammer) is generating a signal (called the jamming signal). This interrupt affects the quality of the signal received and causes several errors, without knowing that the damaged packets are being received and waiting for the receiver to re-transmit them. Second, jamming of the sender is the data link in the receiving object's OSI layer, where the jammer sends a jammer signal to avoid the neighboring objects of their packets from transmitting when you know that the Wi-Fi is busy. Jamming techniques that can be practiced by jammers include many forms of jamming, such as persistent jamming, manipulative jamming, or random jamming. This is a method of detection. Unlike other nodes-centered jam detection solutions, B. Upadhyayas *et al.* [19] demonstrate high detection accuracy without overhead communication costs between the nodes.

7.5.3.2 Vampire Attack

This attack takes advantage of the fact that most IoT artifacts have a small capacity (drummer), when a malicious user misuses devices in such a way as to consume additional power that causes the battery to run out faster, causing service interruption.

The harm caused by the attack is usually determined by the additional energy that artifacts consume when negative activity is not present. Four forms of vampire attack are based on the vacuum technique.

7.5.3.3 Denial of Sleep

In order to minimize the power absorption by intelligent artifacts, different protocols in the data link layer have been suggested by switching these items to sleep where appropriate. The protocols are based on the concept of a system of tax cycles, by exchanging control messages to synchronize objects on their schedules to decide on signal transmits at that interval for the rest of the time. An intruder may now initiate sleep attack denials that do not allow artifacts to switch to sleep through the usual transmission and activation of control signals to adjust their duty cycles. If control messages are encrypted, the attacker will intercept and replay any of the encrypted control messages and force the object nodes to amend their synchronization and scheduling. Mitigation of sleep attacks can be done with encryption, using a time stamp concept or a number in the encrypted message of control, of those control messages which manage and reorganize object node schedules. In ref. [20] K. K. Krentz *et al.* describe three separate DoD attacks, the shooting of ding-dong, crash attacks, and pulse delays, against ContikiMAC.

7.5.3.4 Flooding Attack

The intruder will flood the nearby object nodes with stupid packets and ask them to bring the packages to the fog system where devices collect waste electricity by receiving and sending devices by transmitting such stupid packets. Malicious attackers during UDP flood attacks send the victim a lot of packets, which cause the victim to generate many ICMP messages that block it [21]. Attack mitigation can be accomplished by limiting the packet rate that can be produced by any unit.

7.5.3.5 Carousel Attack

This attack focuses on the network layer of the OSI stack. The routing protocol can be started if it supports source routing. The entire routing path for the packets can be specified when the object node created is routed by the source. In this case, the attacker defines a route that includes closed paths (loops) where the same packet is routed back and forth to exploit its control. The carousel attack shows in Figure 7.3. Carousel connection mitigation can be achieved by using the same source path to delete packets with loops from their routes, as they are most possibly from malicious users by requesting a packet from each object node based on a specified route. As for a carousel attack, it loops a network packet until it eventually reaches its spot. Along with other forms of attack, stretch and carousel attacks can be used. If used alone, however, they can remain undetected, making defense difficult [22].

7.5.3.6 Stretch Attack

The network layer of the OSI stack is targeted by this attack. The routing protocol can send packets to the fog network through very long not direct and short paths if the routing protocol supports source routing. The stretch attack is revealed in Figure 7.4. Mitigation of stretch attacks can be accomplished by preventing source routing, or by

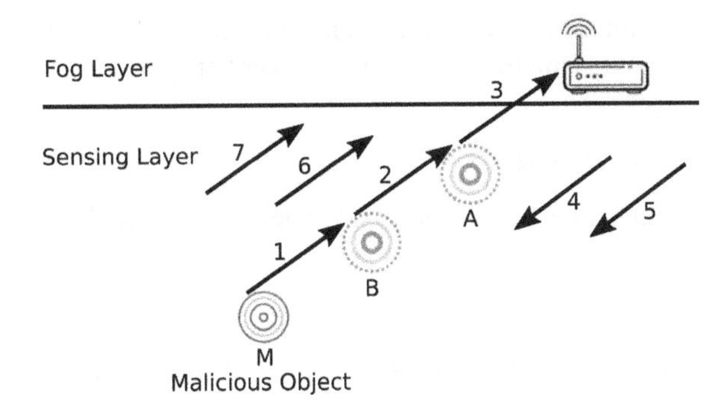

FIGURE 7.3 Carousel attack.

ensuring that redirected packets advance to their targets without long paths. By combining flood attack with carousel attack and stretch assault, the attacker may further increase the amount of waste energy. To this end, an attacker floods the nearby object nodes with many generated packets. He defines long distances loops in order to increase the amount of energy waste.

7.5.3.7 Selective-Forwarding Attack

This type of attack is when the object does not directly send the generated packets to the fog system, but depends on other object nodes. The route to the fog system to deliver those packets includes more object nodes. A disappointed object in this attack does not forward any of the neighboring object packets. The blackhole attack, in which the attacker loses all packets from the night nodes, is a special form of attack. For sensitive IoT applications, packet drops can be avoided by improvements in the transmitting ability of object nodes to enter the fog system directly without any support from intermediate object nodes. However, not every IoT object is supposed to enter the fog directly with high transmission range, and some objects can therefore rely on other objects to deliver their packets, rendering them vulnerable to this attack. A. Mathur *et al.* security mechanism against selective forwarding offenses offers an

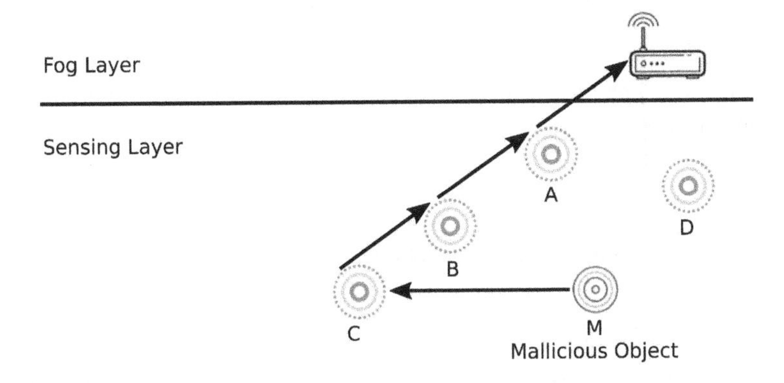

FIGURE 7.4 Stretch attack.

effective way for dealing with single attacks and joint attacks. This can be useful since attacks in the real world sometimes function together between malicious nodes [23].

7.6 SECURITY REQUIREMENTS

The safety requirements for IoT are included under A. Rayes and S. Salam [17].

7.6.1 Confidentiality

This ensures that only the dedicated and intended entities can understand the messages to be exchanged between devices.

7.6.2 Integrity

This means that a third party or other entity does not adjust, alter, or dampen the messages to be exchanged.

7.6.3 Authentication

It means that the individuals involved in every project claim to be who they are. This law is generally taken into account when a masquerade attack or an immigrant attack assumes a different identity.

7.6.4 Availability

This implies a continuous service. Denial of service attacks aim at this situation because it causes service harm and disruption.

7.6.5 Authorization

This ensures that businesses have the requisite control and access authorizations for carrying out the activity they need.

7.6.6 Freshness

The data is revised and refreshed. Replay attacks take advantage of this situation to render an object older if an old message is being replayed.

7.6.7 Non-Repudiation

This means that an entity's activity is not rejected.

7.6.8 Forward Secrecy

It means that when an entity enters the network, the correspondence exchanged after its departure cannot be understood.

7.6.9 Backward Secrecy

This means that any new entity connecting to the network cannot understand the correspondence exchanged before entering the network.

7.7 IoT SECURITY CHALLENGES

Some protection problems are required to be incorporated with IoT, according to Roman *et al.* [14].

7.7.1 Protocol and Network Security

The heterogeneity of the IoT protocol and network security infrastructure has a huge influence. Limited devices communicate either via gateways or directly with multiple heterogeneous devices (e.g., other restricted devices, full-blown web servers). In this case, efficient cryptographic algorithms can be implemented that provide a high output independent from their design (same 8-bit or 16-bit devices), and lightweight security protocols can be acceptable or developed, which provide secure communication between devices. In this case, it is also important. These protocols require qualities or credentials, so that the required session keys are defined among device pairs with optimal key management systems to enforce and distribute these credentials.

7.7.2 Identity Management

The existence of billions of heterogeneous objects can have an impact on identity management. Some methods for universal authentication include the proper identity. Without authentication, it is not possible to ensure that a specific entity's generated data flow contained what should be contained. Authorization is another important aspect. If there is no control of access, then everything that is neither realistic nor viable will be accessed by everyone.

7.7.3 Privacy

Actually, a significant flood of data generated by thousands of information creators poses an immense privacy threat. Users should have tools in this closely interconnected world that allow them to maintain their impersonality. Some instruments should provide an instant picture of a particular user's policies and records, allow accountability, and prevent IoT from secretly controlling our lives. The IoT itself must therefore specifically take care of the application of privacy by some design standards and must have user-centric protection and privacy support. Centric confidentiality, content-oriented data protection, or context-oriented privacy are supported by existing privacy frameworks. However, IoT networks naturally include autonomous nodes, which collect data and require object-oriented data protection modeling [24].

7.7.4 Trust and Governance

The IoT network's size and complexity will affect its confidence and governance. Two dimensions of confidence can be taken to mean: (a) trust in the interaction between devices and entities, where uncertainty is addressed concerning the future actions of all collaborating devices and entities, and (b) user confidence in the system, as users must be in a position to manage their devices and things in order to feel free and independent. As for governance, it is like an advantage on both sides, which should be carefully practiced. It should, on the one hand, offer political decisions such as support, stability, and the ability to define common frameworks and mechanisms of interoperability. On the other hand, governance can easily be over-controlled and promote an over-controlled climate.

7.7.5 Fault Tolerance

In the sense of IoT, the number of attack vectors and device vulnerability considerations will undoubtedly increase, rendering fault tolerance an integral feature. We will need the creation of awareness mechanisms to build foundations of intrusion detection and prevention mechanisms alongside the attempt to achieve protection by default (robust implementations, functional systems, etc.) in IoT. This will help to protect IoT entities and devices. Finally, recovery services should be considered, which should be able to locate insecure areas and entities (i.e., attack areas) and redirect systems' functionality to other trusted states and areas.

7.8 PRIVACY AND TRUST

In Lessig, confidentiality is described as the combination of "controlling power," "security utility," and "balancing integrity" and "regulating power balance agent." Many meanings categorize data security as a prerequisite for PII, e.g. "personally identifiable information (PII) leaves the person whose PII it is under control [25]."

J. Daubert *et al.* [25] indicates that privacy can be classified into different dimensions, such as the protection of identity, privacy of the venue, privacy of the footprint, and privacy of queries. The privacy of identity is the need for privacy in order to identify a person by pseudonymizing or anonymizing the PII. Privacy of locations refers to an exploratory and specific form of privacy on the footprint. The position may signify several fold PIIs, such as the church location of interest and religion.

The need for location protection solutions can be enforced, such as exchanging location slices. The protection of Footprint relates to the need for protection for all accidentally leaked PIIs, such as preferred language and the web-based operating system. This data is also called micro and metadata. The fact that the action is carried out alone can leave the footprint unimpeded. The privacy provision for PII data included in a request applies to the need for anonymity, e.g., the location and date of weather forecasts. Question response approaches are available that respect the privacy requirement. As per [26], new IoT data collection methods have created many problems for privacy. Some of the challenges are to obtain a data collection agreement that allows users to track, configure, and select the data they exchange, and to ensure that the data collected is used exclusively for the specified purpose. The increased abuse of personal information in the IoT domain makes these challenges more difficult. This comes from the long-term monitoring of habits, behaviors, and locations. The concept of trust varies in different contexts and different meanings, according to S. Sicari at the El [1]. Confidence is a complex concept that in the scientific literature there is no definite harmony with. It is, however, widely recognized for its importance.

The problem with many approaches to the concept of trust is that benchmarks and assessment methodologies are not being developed. Ignore nodes in an IoT network are meant by a confidence-based attack, such as self-promoting, bad mouthing, and good mouth breaking the fundamental functionality of IoT. Confidence assessment criteria are: reliability, collaboration, and the welfare of the group. A dynamic trust management protocol is thus able to adjust to dynamically changing environments and can use the best confidence parameter configuration to optimize application efficiency. The credibility of social networking principles into IoT can be tested in Social Internet of Things (SIoT). As IoT infrastructure artifacts can establish social relationships with their owners in an autonomous way, trust can be evaluated.

There are four trust dimensions in J. Daubert *et al.* [25], namely device trust, service provider trust, connection trust, and system trust (overactive). Device confidence refers to the need for reliable devices like sensors and actuators to interact. Confident computing and trusted software are growing approaches to achieving this aim. Trust in processing means that accurate and relevant data need to be handled. The accurate collection of data, combined with adequate analysis, will accomplish this. Data fusion will further boost the performance. Connection trust means that the right and correct data must be exchanged with and only with right and genuine service providers. Canonical security goals may be met,such as confidentiality, honesty, and non-repudiation. The ability to help a secure overall system implies the confidence of the system. This level of trust can be accomplished by providing all aspects of workflows, procedures, underlying technologies, etc., and transparency.

7.9 IoT SECURITY MECHANISM

M. Mohamad Noor and W. Hassan discussed [27] current IoT security mechanisms. Their review considered the authentication, encryption, confidence management, and safe routing of data, infrastructure, and device protection, and the availability of IoT ecosystem services, as protection mechanistic and technologically new ones such as SDN (Software Defined Network) and Block Chain.

7.9.1 Authentication

The available authentications are a specific process, and the use of a public key for easy authentication is still not the best choice to reduce security, since you can steal a public key. Any malicious codes or comments can also bypass authentication. Current IoT authentication solutions may have vulnerabilities.

7.9.2 Encryption

The nodes must be encrypted to achieve end-to-end safety. Due to the heterogeneity of the IoT systems, however, some nodes may incorporate microprocessors for the general purpose. Only application-specific ICs can be embedded in low resources and restricted tools. The aim is to reach efficient end-to-end communication, low power consumption, symmetrical, and asymmetrical lightweight algorithms for IoT that can meet these requirements. Lightweight encryption is only an efficient encryption system for such instruments.

7.9.3 Trust Management

IoT Trust Management's main purpose is to detect, remove, and secure access control nodes. The trust management research stands out for the automatic and dynamic calculations to validate the trust values of the participating instrument and nodes in an IoT network. Due to scalability and the enormous number of clever items containing sensitive data, an automated, transparent, and simple access control management is urgently needed so that different node/user access rates can be provided. The on-off attack that affects the trust worth of the node should be mitigated through a smart trust evaluation using Machine Learning (ML). A generic confidence computer model and a feature removal method which may be applied in any IoT service scenario, using a method to label the data according to its confidentiality [28]. Therefore, the trust management must be able to balance the obvious authentication vulnerability, such as attacks from the nodes affected and malicious.

7.9.4 Secure Routing

For the secure, autonomous, and energy-efficient routing scalability of any routing solution, the LLN connects to the internet or closes via the Local Area Network (LAN), some of the sensor nodes in the IoT network. The IP addresses for these devices can be based on IPv6, considering the huge size of the IoT networks. The RPL is for multi-point communication that fulfills all of LLN's routing requirements; it is vulnerable to many security attacks. A reliable and efficient protocol must facilitate mutual authentication, key exchange, complete confidentiality, and protected routing privacy.

7.9.5 New Technologies

Latest IoT protection solutions are emerging technologies like SDN and Block Chain. The key idea of SDN is to divide the control of the network and the data. The crypto-monetary block chain is the backbone. IoT-based applications can take advantage of their secure and private transactions and their process and communications decentralization. Its application in financial applications has so far achieved considerable success. The main elements of BC are transactions generated by the participants and recording blocks. Here, the block of the recording system checks whether or not the transaction information has been kept in the correct order. This does not allow the data available to be manipulated. This makes secure authentication possible [29]. In addition to this, certain current IoT challenges, such as reliability, safety, scalability, and yet efficiency, must be addressed.

REFERENCES

1. S. Sicari, A. Rizzardi, L. A. Grieco, and A. Coen-Porisini, "Security, privacy and trust in internet of things: The road ahead," *Comput. Networks*, vol. 76, pp. 146–164, 2015.
2. I. Bhardwaj, A. Kumar, and M. Bansal, "A review on lightweight cryptography algorithms for data security and authentication in IoTs," *4th IEEE Int. Conf. Signal Process. Comput. Control (ISPCC 2017)*, January 2017, pp. 504–509, 2017.
3. S. Hameed, F. I. Khan, and B. Hameed, "Understanding security requirements and challenges in internet of things (IoT): A Review," *J. Comput. Networks Commun.*, vol. 2019, pp. 1–14, 2019. https://doi.org/10.1155/2019/9629381
4. C. W. Tsai, C. F. Lai, and A. V. Vasilakos, "Future internet of things: Open issues and challenges," *Wirel. Networks*, vol. 20, no. 8, pp. 2201–2217, 2014.
5. W. Zhou, Y. Jia, A. Peng, Y. Zhang, and P. Liu, "The effect of IoT new features on security and privacy : New threats, existing solutions, and challenges yet to be solved," *IEEE Internet Things J.*, vol. 6, no. 2, pp. 1606–1616, 2019.
6. F. Idris and S. Hameed, "Software defined security service provisioning framework for internet of things," *Int. J. Adv. Comput. Sci. Appl.*, vol. 7, no. 12, pp. 1–15, 2016.
7. K. T. Nguyen, M. Laurent, and N. Oualha, "Survey on secure communication protocols for the internet of things," *Ad Hoc Networks*, vol. 32, no. 2015, pp. 17–31, 2015.
8. W. Meng, "Intrusion detection in the Era of IoT: Building trust via traffic filtering and sampling," *Computer (Long Beach. California)*, vol. 51, no. 7, pp. 36–43, 2018.
9. G. Sun *et al.*, "Efficient location privacy algorithm for internet of things (IoT) services and applications," *J. Netw. Comput. Appl.*, vol. 89, pp. 3–13, 2017.
10. C. Ioannou and V. Vassiliou, "An intrusion detection system for constrained WSN and IoT nodes based on binary logistic regression," *21st ACM Int. Conf.*, pp. 259–263, 2018.
11. S. Ezekiel, D. M. Divakaran, and M. Gurusamy, "Dynamic attack mitigation using SDN," *2017 27th Int. Telecommun. Networks Appl. Conf. ITNAC 2017*, January, pp. 1–6, 2017.
12. M. A. Razzaque, M. Milojevic-Jevric, A. Palade, and S. Cla, "Middleware for internet of things: A survey," *IEEE Internet Things J.*, vol. 3, no. 1, pp. 70–95, 2016.
13. B. Cusack, Z. Tian, and A. K. Kyaw, "Identifying DOS and DDOS attack origin: IP traceback methods comparison and evaluation for IoT," *Lect. Notes Inst. Comput. Sci. Soc. Telecommun. Eng. LNICST*, vol. 190, pp. 127–138, 2017.
14. R. Roman, J. Zhou, and J. Lopez, "On the features and challenges of security and privacy in distributed internet of things," *Comput. Networks*, vol. 57, no. 10, pp. 2266–2279, 2013.
15. K. M. Sadique, R. Rahmani, and P. Johannesson, "Towards security on internet of things: Applications and challenges in technology," *Procedia Comput. Sci.*, vol. 141, pp. 199–206, 2018.
16. J. Ahamed and A. V. Rajan, "Internet of things (IoT): Application systems and security vulnerabilities," *Int. Conf. Electron. Devices, Syst. Appl.*, pp. 1–5, 2017.
17. A. Rayes and S. Salam, *Internet Things From Hype to Reality: The Road to Digitization*, Springer, Berlin, Germany, 22nd Oct, 2016, 2019.
18. M. Nawir, A. Amir, N. Yaakob, and O. B. Lynn, "Internet of things (IoT): Taxonomy of security attacks," *2016 3rd Int. Conf. Electron. Des. ICED 2016*, pp. 321–326, 2017.
19. B. Upadhyaya, S. Sun, and B. Sikdar, "Machine learning-based jamming detection in wireless IoT networks," *Proc. IEEE VTS Asia Pacific Wirel. Commun. Symp. APWCS 2019*, pp. 1–5, 2019.

20. K. Krentz, C. Meinel, and H. Graupner, "Countering three denial-of-sleep attacks on ContikiMAC," *EWSN*, February, pp. 108–119, 2017.
21. Kamaldeep, M. Malik, and M. Dutta, "Contiki-based mitigation of UDP flooding attacks in the Internet of things," *Proceedings IEEE Int. Conf. Comput. Commun. Autom. ICCCA 2017*, January, pp. 1296–1300, 2017.
22. R. Smith, D. Palin, P. P. Ioulianou, V. G. Vassilakis, and S. F. Shahandashti, "Battery draining attacks against edge computing nodes in IoT networks," *Cyber-Physical Syst.*, pp. 96–116, 2020. https://doi.org/10.1080/23335777.2020.1716268
23. A. Mathur, T. Newe, and M. Rao, "Defence against black hole and selective forwarding attacks for medical WSNs in the IoT," *Sensors (Switzerland)*, vol. 16, no. 1, 2016.
24. M. Conti, A. Dehghantanha, K. Franke, and S. Watson, "Internet of things security and forensics: Challenges and opportunities," *Futur. Gener. Comput. Syst.*, vol. 78, pp. 544–546, 2018.
25. J. Daubert, A. Wiesmaier, and P. Kikiras, "A view on privacy & trust in IoT," *2015 IEEE Int. Conf. Commun. Work. ICCW 2015*, pp. 2665–2670, 2015.
26. P. Emami-Naeini *et al.*, "Privacy expectations and preferences in an IoT world," *Proc. 13th Symp. Usable Priv. Secur. SOUPS 2017*, pp. 399–412, 2019.
27. M. Binti Mohamad Noor and W. H. Hassan, "Current research on internet of things (IoT) security: A survey," *Comput. Networks*, vol. 148, pp. 283–294, 2019.
28. U. Jayasinghe, G. M. Lee, T. W. Um, and Q. Shi, "Machine learning based trust computational model for IoT services," *IEEE Trans. Sustain. Comput.*, vol. 4, no. 1, pp. 39–52, 2019.
29. N. M. Kumar and P. K. Mallick, "Blockchain technology for security issues and challenges in IoT," *Procedia Comput. Sci.*, vol. 132, pp. 1815–1823, 2018.

8 A Framework for Delivering IoT Services with Virtual Sensors

Case Study Remote Healthcare Delivery

Nandini Mukherjee, Sunanda Bose, and Himadri Sekhar Ray
Jadavpur University, Kolkata, India

CONTENTS

8.1 Introduction 137
8.2 Related Work 138
8.3 System Overview 140
8.4 Virtual Sensors and SensIaas 141
8.5 Kiosk-Based Healthcare Delivery 144
8.5.1 KiORH Application—A Case Study 145
8.5.2 Features of KiORH 146
8.6 Impact on The Society 149
8.7 Conclusion 150
References 150

8.1 INTRODUCTION

Internet of Things (IoT) creates a new paradigm of emerging applications, which were never thought of before. Starting with simple applications like "environment monitoring," "habitat monitoring," IoT services can be developed for complex applications like "industrial automation," "smart cities," and "healthcare." Cloud and sensing technologies are the two primary ingredients for building any IoT platforms. However, while integrating the sensor devices with the cloud environment, several challenges are faced by the developers. Firstly, heterogeneous devices are available in the market and they use heterogeneous communication interfaces and protocols. Therefore, handling these devices from a cloud-based application requires the use of hardware-specific interfaces, curbing interoperability and portability. Secondly, sharing of these sensor outputs by various applications in real time is practically impossible. There are other issues as well, like "how to handle the uncertainty factors in the data," and "how to store the unstructured and semi-structured data," etc.

Sensor virtualization is a technique that creates a software abstraction of the physical sensors. We propose to create different categories of virtual sensors, based on the application requirement. We also propose a framework based on these virtual sensors for handling the heterogeneity, interoperability, and portability issues of physical sensors. The framework allows different applications to share the sensor output in real time. It also takes care of storage issues for unstructured and semi-structured data.

In this chapter, we focus particularly on remote healthcare service delivery. Providing basic healthcare services in rural areas of developing countries is a challenge. The basic healthcare facilities and doctors are not available in the same proportion in rural areas as they are available in urban areas. Some studies in India indicate that there are about four times as many trained doctors per ten thousand population in urban areas as compared to the rural areas. This includes doctors in the public sector (in primary healthcare centers set up by the government), as well as in private hospitals and nursing homes and privately practicing doctors. Clinical testing facilities are also unavailable in rural areas. The goal of this research work is to increase the reachability of rural people to healthcare services with the use of the proposed IoT framework.

The focus is on primary healthcare delivery, and two basic use cases are considered:

- Use-Case 1: Continuous monitoring of elderly patients or patients on the move
- Use-Case 2: Kiosk-based healthcare delivery

In order to demonstrate how the current state-of-the-art technologies can be used for providing efficient solutions in the above cases, prototype applications have been developed. A brief discussion on one of the applications is discussed in this chapter.

Organization of the chapter is as follows. Related work is presented in Section 8.2. Section 8.3 gives an overview of our proposed IoT framework. Section 8.4 introduces virtual sensors and briefly describes the SensIaaS architecture for hosting the virtual sensors. Section 8.5 discusses a kiosk-based remote healthcare delivery application. The impact of the application on the society is discussed in Section 8.6, and finally the chapter is concluded in Section 8.7.

8.2 RELATED WORK

The concept of virtual sensors is in use during the last two decades. In 2002, Albertos *et al.* proposed to use virtual sensors for control applications [1]. They suggested using AI techniques on top of sensing devices to convert a variable into another to perform intelligent tasks. Virtual nodes have been used in [2] for simplifying the development of decentralized WSN applications. In this research work, the data acquired by a set of sensors are collected and processed according to an aggregation function which is provided by the application. The set of physical nodes are abstracted into a virtual one, using logical neighborhoods as discussed in [3]. In [4], virtual sensors have been used to create abstractions of components of body sensor networks, and a multi-layer task model is presented based on the concept of virtual sensors to improve architecture modularity and design reusability. The authors propose an open source domain-specific framework, SPINE2, designed to support distributed sensing operations and signal processing for wireless sensor networks and to enable code reusability, efficiency, and application interoperability.

While the above works focused on creating virtual sensors on top of a set of physical sensors, in [5], sensed data from physical sensors and data computed by virtual sensors are combined to estimate an external wrench acting along the structure of a robot manipulator. The authors propose to use a torque sensor mounted at the robot base, and results from a model-based virtual sensor, and integrate the two types of measurements to estimate the unknown quantities efficiently. On the other hand, in [6], virtual sensing is used to give an estimate of a quantity of interest using the measurements from an AB array of physical sensors, using hardware redundancy. The authors of this research work proposed structural health monitoring of large structures, using an array of sensors for vibration measurements and Bayesian virtual sensors for better accuracy. A virtual sensor architecture is used in [7] to reconstruct an unmeasurable scheduling signal of a parameter-varying system from input and output measurements. Here, an Artificial Neural Network is used with input/output measurements and with data generated by processing such measurements through a bank of linear observers. This approach is used for fault detection and isolation, predictive maintenance, and gain-scheduling control.

In our research work, we use the first approach, that is, virtual sensors are deployed as software abstractions on top of physical sensors. However, unlike the previous approaches, we classify the virtual sensors into different categories based on the application requirements.

The framework for management of virtual sensors is another research area, which has been explored by many researchers. Sarkar *et al.* [8] presents a virtual sensing framework (VSF). The objective of this framework has been to reduce sensing and data transmission activities of physical sensor devices. The framework creates virtual sensors at the sink to exploit the temporal and spatial correlations among sensed data. They use an adaptive model on top of virtual sensors to predict consecutive sensed data for all the nodes, with the help of sensed data from a few active nodes. In [9], the authors deal with virtual sensor provisioning based on the similarity of heterogeneous sensors, with an objective of reducing energy consumption. Ojha *et al.* in [10] present a scheme for dynamic virtual sensor provisioning for IoT applications to reduce energy consumption of the physical sensor nodes, and at the same time, to maintain the quality of services. They proposed a model for the interaction between the cloud service providers and the sensor owners, using the single-leader multifollower Stackelberg game.

We present a framework for efficient creation and provisioning of virtual sensors based on application requirements and allocation of physical sensors to the virtual sensors to hide the heterogeneity of hardware and communication standards from the application developers.

The application domain selected for our research work is healthcare. IoT-enabled healthcare applications and services have been proposed by several researchers [11–14]. The authors in [11] deal with the security and integrity of IoT-enabled healthcare data. They propose a hybrid security model for securing the diagnostic text data in medical images. In [12], a wearable sensor node with solar energy harvesting and Bluetooth low-energy transmission is used that enables implementation of an autonomous WBAN. Sensor nodes are deployed on the body to measure body temperature, heartbeat of the patient, and to detect falls. A Fog-based IoT healthcare solution structure is proposed in [13]. A virtual sensor that computes the blood pressure of patients using a limited number of physical sensors has been used in [14]. The sensor data are ported to the cloud and used as the inputs of virtual sensor.

In the next sections of this chapter, we present a virtual sensor-based framework for building complex IoT applications. Within this framework, the virtual sensors are treated as the members of the shared resource pool, just like other resources in the cloud. We propose an architecture for scheduling and provisioning virtual sensors at the IaaS layer of the cloud. Finally, we discuss an application for remote healthcare service delivery based on our framework.

8.3 SYSTEM OVERVIEW

Figure 8.1 shows a layered architecture of our framework. The bottom three layers of the framework are generic and can be used by any IoT application. The lowest layer consists of hardware devices, including different types of stand-alone sensors, wireless sensor network, body sensor network, and storage and computing devices. On top of these hardware devices, a secure sensor cloud framework is deployed, which offers services like addition, removal, reservation, and allocation of resources. Here, resources imply both sensing and computing resources. However, the resources are virtualized before being accessed by the upper layers. This layer, together with the set of basic services, constitutes the IaaS layer of our sensor-cloud framework. A set of storage services can also be built on top of this layer. Data models and meta-information are used with these services to design the storage templates for specific applications.

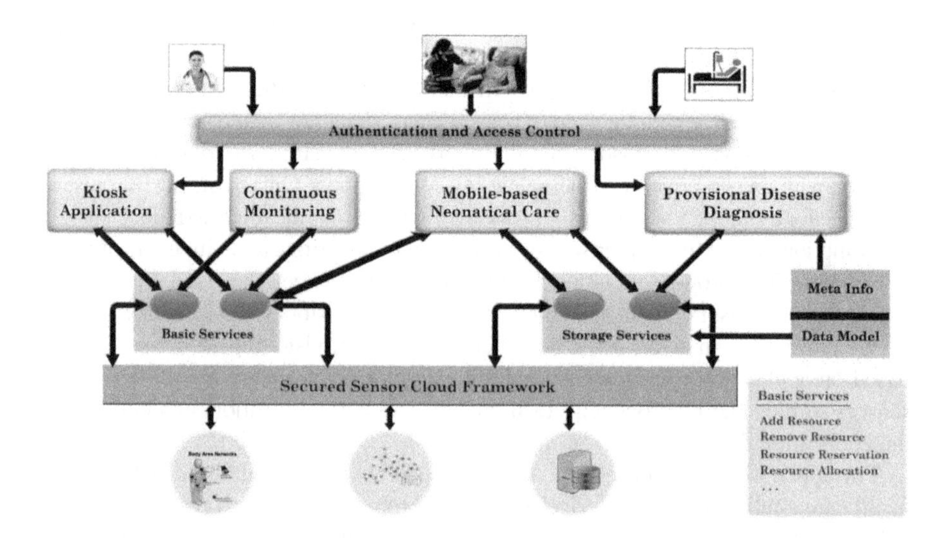

FIGURE 8.1 Layered architecture.

On top of the sensor-cloud layer, different applications are built. However, these applications are independent of the underlying hardware, as the sensor-cloud layer handles all the complexities of heterogeneous hardware devices and communication standards and protocols. The sensor-cloud layer consists of virtual sensors, which offer on-demand, real-time, and application-specific sensing services to the application developers. The virtual sensors are exposed to the applications at the PaaS or SaaS layers, just like other virtualized resources in the cloud.

In the next section, the details of different types of virtual sensors are presented. An architecture is also discussed for our sensor-cloud layer, which handles the virtual sensors along with other virtualized resources.

8.4 VIRTUAL SENSORS AND SENSIAAS

It is understood that varying sensing requirements may arise for different scenarios for providing IoT services. In the healthcare domain, continuous monitoring of body vitals of remotely located patients may be monitored by the doctors. Often, doctors located in urban areas need to treat the rural patients remotely, due to the patients' inability to travel long distances in a short time or frequently. The hospital authorities may use such systems to monitor a group of patients and provide insightful analytics. This requires fusion and sharing of sensing resources to multiple sensor-aware consumer applications simultaneously. For the sensor-aware applications, it is necessary to provide uniform access to the remotely located streaming devices in order to felicitate the sensing services. Virtual sensors abstract one or more physical sensors and provide a uniform access layer. There can be various types of mapping between the physical sensors and the virtual sensors, including one-to-one, one-to-many, many-to-one, and many-to-many, as shown in Figure 8.2. When multiple physical sensors are abstracted with a single virtual sensor, then there can be various accumulation and aggregation strategies applied on the data received from the physical sensors. When the same physical sensor is shared by multiple virtual sensors, then proper sharing techniques are required.

Depending on the usage, virtual sensors may be categorized into different types as mentioned below.

Singular: A single physical sensor is interfaced with a virtual sensor. This type of virtual sensors corresponds to one-to-one mapping between the physical and virtual sensor. This type of virtual sensors can be used in IoT-based healthcare monitoring, where a single sensor, measuring the body vitals of a patient, is interfaced with a virtual sensor.

$$r_i \rightarrow v_i \qquad \left.\begin{matrix} r_1 \\ \cdots \\ r_n \end{matrix}\right\} \rightarrow v_i \qquad r_i \rightarrow \left\{\begin{matrix} v_1 \\ \cdots \\ v_n \end{matrix}\right. \qquad \left.\begin{matrix} r_1 \\ \cdots \\ r_n \end{matrix}\right\} \rightarrow \left\{\begin{matrix} v_1 \\ \cdots \\ v_m \end{matrix}\right.$$

(a) one-to-one **(b) many-to-one** **(c) one-to-many** **(d) many-to-many**

FIGURE 8.2 Mapping between physical and virtual sensors.

Selector: A subset of physical sensors is represented by a single virtual sensor by selecting one at any instance of time. This type of virtual sensors corresponds to many-to-one mapping between physical and virtual sensors. This type of virtual sensors can be used for IoT-based environment monitoring, where a group of homogeneous sensors (e.g., a set of temperature sensors located in a room) are represented as a single virtual sensor. At any instance of time, one of the available physical sensors is used to minimize energy consumption.

Accumulator: Like selectors, a subset of physical sensors is represented by a single virtual sensor. However, instead of selecting a single physical sensor, accumulation of all sensors in that subset is represented as the virtual sensor. This type of virtual sensor corresponds to many-to-one mapping between physical and virtual sensors. Sensed traffic of all the physical sensors in the subset is combined and fed to the virtual sensor. These virtual sensors are used when monitoring multiple parameters through a set of heterogeneous sensors applicable to both environment monitoring (e.g., temperature, humidity, and light sensor in a room), as well as healthcare monitoring (e.g., pulse, breathing rate, and SpO2 sensors attached with a person).

Aggregator: Similar to the accumulator, but in this case the combined traffic is aggregated by an aggregator function that accepts all sensors as input of that function and returns the result as output. The returned output is relayed to the virtual sensor. This type of virtual sensors are often used in IoT-based environment monitoring, when the average, maximum, or minimum of a set of homogeneous sensors are monitored continuously (e.g., average room temperature is monitored through multiple temperature sensors available in the same room).

Qualifier: A qualifier is much like a singular virtual sensor; however, it remains idle unless the sensed data from the physical sensor qualifies a given function. Similar to singular, this type of virtual sensor also corresponds to one-to-one mapping. These sensors may be used when monitoring is not required, unless some special criterion is met (e.g., monitor body temperature only when it gets higher than 37°C).

Context qualifier: Context qualifiers also use a qualifying function. However, instead of applying the function on a single sensor, it is applied on a subset of sensors. The qualifier function returns a subset of qualified sensors. The traffic from the qualified sensors are combined and relayed to the virtual sensor. This type of virtual sensors also corresponds to many-to-one mapping between the physical and virtual sensors. These sensors are used when either a single parameter is not enough to qualify as certain criteria for sensing, or it is required to observe multiple sensing parameters (e.g., monitor breathing rate and pulse when temperature exceeds 37°C).

Some use cases of virtual sensing in healthcare domain are presented in Figure 8.3.

The virtual sensors depicted in the figure are described below.

- Different health sensors (SpO2 (1), pulse (2), and temperature (3)) are attached on a patient's body. Data from these sensors are accumulated to one virtual sensor. Semantics of such fusion may be defined by the sensor-aware application, using that virtual sensor which may be useful to its end users, e.g., doctors.

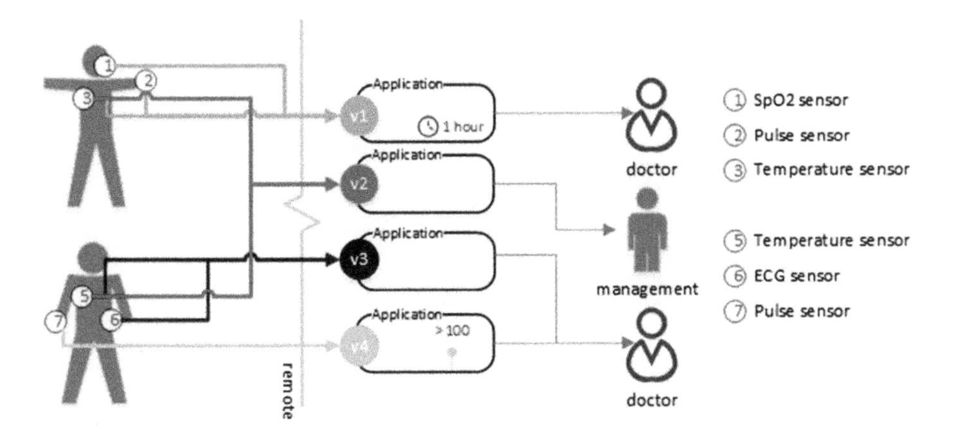

FIGURE 8.3 Use cases of virtual sensors.

- A set of sensors measuring a common clinical parameter (temperature (3, 5)) of all patients is requested by an application, which may be used by the management authority for continuous monitoring of health status of all patients.
- Different sensors attached on the body of another patient (temperature (5), ECG (6)) may be requested by another sensor-aware application.
- Some sensors (pulse sensor (7)) may be requested to check if it crosses a threshold value. It may be used to raise an alarm or trigger some event.

In order to deploy healthcare services and run them efficiently, we developed a sensor-cloud architecture that enables virtual sensors, i.e., a layer of abstraction at the IaaS (Infrastructure as a Service) level [15]. Thus, remotely located sensing resources are treated as shared resources in a resource pool similar to CPU, memory, and storage devices. In most of the contemporary research works, integration of sensors in a cloud environment is done at the application layer. On the other hand, in our architecture, virtual machines (VM), which are hosted in the physical machines (PM) on top of the hypervisor, are attached with virtual devices, each corresponding to one or more physical sensors, transmitting remotely over a many-to-many connection. The goals achieved with this approach are as follows: (i) heterogeneity of the resources (here sensing devices) are handled; (ii) no special resource management techniques are required and generic management policies can be developed for both, computational as well as sensing resources; (iii) additional capabilities can be added to the virtual sensors that help to overcome the resource limitations of physical sensors; and finally, (iv) resource demands made from applications which include overlapping subsets of sensing resources having different semantics can be efficiently handled using virtual sensors. Moreover, sharing a single sensing resource by multiple consumer applications over a many-to-many connection can also be incorporated. This sharing is particularly useful for continuous monitoring of patients (i.e., *Use Case-1*) applications.

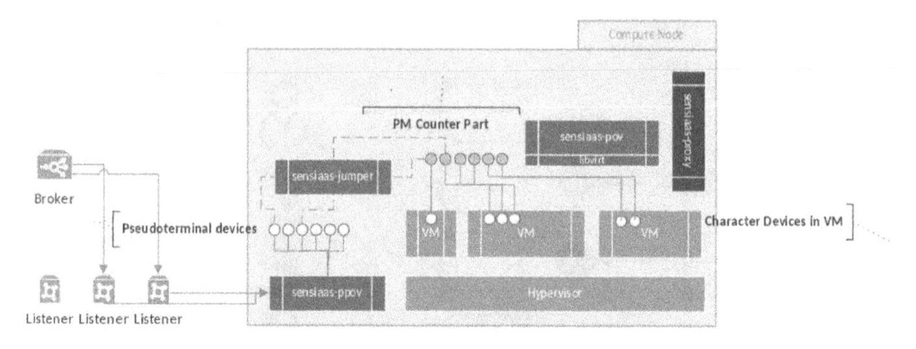

FIGURE 8.4 SensIaaS architecture.

The proposed Sensor Cloud Infrastructure spans over multiple geographic boundaries. The *Resource Hosts* located remotely host the sensing devices. A *Cloud Station* located at the data center is supposed to deliver virtual machines with virtual sensing abilities. A sensing device inside the resource host may comprise multiple sensors of different types. A subset of those sensors may be acquired to satisfy multiple sensing services. So the sensing device may transmit the sensed data of the selected subset to the cloud station. In the cloud station, that sensed data is de-multiplexed and distributed to the appropriate virtual machines. The cloud station consists of different components, synchronizing in order to provisioning of remotely located sensing resources. Figure 8.4 provides an overview of the architecture. *Sensiaas-asset* is the component responsible for maintaining the inventory of resources and the allocation information. *Sensiaas-asset* also includes a *Resource Broker* that performs the resource selection task. *Resource Listeners* are responsible for listening to the incoming sensed traffic, and they forward that to an appropriate physical machine, which hosts the virtual machines that are associated with that stream. Multiple *Resource Listeners* may be load-balanced by a *Listener Broker.* Once the sensed traffic reaches a physical machine, it is demultiplexed and the sensed data of individual sensors are stripped into different streaming pseudo devices by *PPoV (Primary Point of Virtualization).* These pseudo devices are termed as PPoV endpoints. On the other hand, the virtual machines are equipped with pseudo devices termed as *PoV (Point of Virtualization)* that can be read by the sensor-aware applications hosted inside. Another component, known as *Jumper,* connects one or more PPoV endpoints with its corresponding PoV endpoint. The PPoV endpoints may be shared across multiple PoV endpoints to enable sharing of remotely located sensing resources.

8.5 KIOSK-BASED HEALTHCARE DELIVERY

We have experimented with a kiosk-based healthcare delivery application on top of our SensIaaS architecture. To overcome the problem of providing the primary healthcare services in the rural areas, multiple kiosk-based health centers can be set up in rural and remote villages using android mobile phones, tabs, and medical

FIGURE 8.5 Overview of KiORH.

sensor devices. A real-time application, known as KiORH (Kiosk Operated Rural Health), is developed to provide an integrated platform to the patient and the urban-located doctors for creating a treatment episode for delivering basic health-care services. The kiosks are supported by the health assistants, who collect the patients' complaint-related data with the help of the android application. This application transmits the data to the doctor through the internet. The doctors sitting at an urban location can go through real-time patient data. The doctor then generates advice, prescriptions, and suggests tests, etc., from the other side of the KiORH application. An overview of the applications is given in Figure 8.5.

8.5.1 KiORH Application—A Case Study

The application KiORH is developed for rural kiosks and the areas where resources like the internet and electricity are big constraints. In general, the kiosks are operated by the health assistants, who perform different tasks like patient registration, measuring vitals of the patients using e-health sensors, asking medical histories, collecting patients' complaints along with symptoms, and then connecting to the doctors for remote consultation with the help of our application. The application has two sections; one is for the kiosk side, and the other section is for the doctor side. The kiosk side application is developed using HTML5, CSS3, JQuery3.x wrapping, with Cordova wrapping in the native container. This native container can access device functions for various platforms. For the doctor section, a web-based application is developed using Python, Django, and PostgreSQL. The health assistants of the rural kiosks have been trained to use e-health sensing devices and to collect the health information of the patient. They have been provided with android mobile tablets with the pre-installed KiORH app. On the other hand, doctors, using the other side of the applications, can check the symptoms and complaint of the patient, along with the details like vitals, history, and even the old prescription, and present old medical reports remotely.

8.5.2 Features of KiORH

The KiORH works differently from other health-related mobile apps and is also suitable for running in the rural areas of the developing countries, where scarcity of resources is a challenge. Some of the important features of the application are described below.

- **Role-based login**: This module is used for user authentication. Through this module, health assistants, doctors, or patients can log in to the system and can access the different modules of the application according to the authority provided for them.
- **Sensor data acquisition:** Sensor data gathering is another important activity of the kiosk application for a real-time flow of sensor data from kiosk to doctor end. There is an e-health shield developed by cooking hacks, which can be equipped with the Arduino board. The sensors, like body temperature sensor, SPO2 sensor, airflow sensor, blood pressure sensor, and body posture sensor are attached with that e-health shield. The sensors sense the vitals and send the data to the attached Arduino board. Then the data is sent to our SensIaaS layer.
- **Symptom collection**: The application uses a knowledge base to collect the symptoms of the patients. The symptoms in the knowledge base are arranged as a set of tree structures rooted at few main symptoms. The health assistant can select one or more main symptoms by tapping the application screen after getting the patient complaints. Main symptoms contain multiple sub-symptoms and the sub-symptoms contain multiple questions and answers related to the currently selected sub-symptom.

  An example of how a health assistant collects symptoms for a patient with headache using the knowledge base is shown in Figure 8.6. At first, from the list of symptoms the health assistant selects *Pain*. Then the list of sub-symptoms of pain will show up and the health assistant selects *Headache* as sub-symptom. Then symptom-related questions like duration, starting location, current location, intensity, nature, etc., and the multiple-choice answers to each question will be displayed. The health assistant selects the answers based on the patient's response.

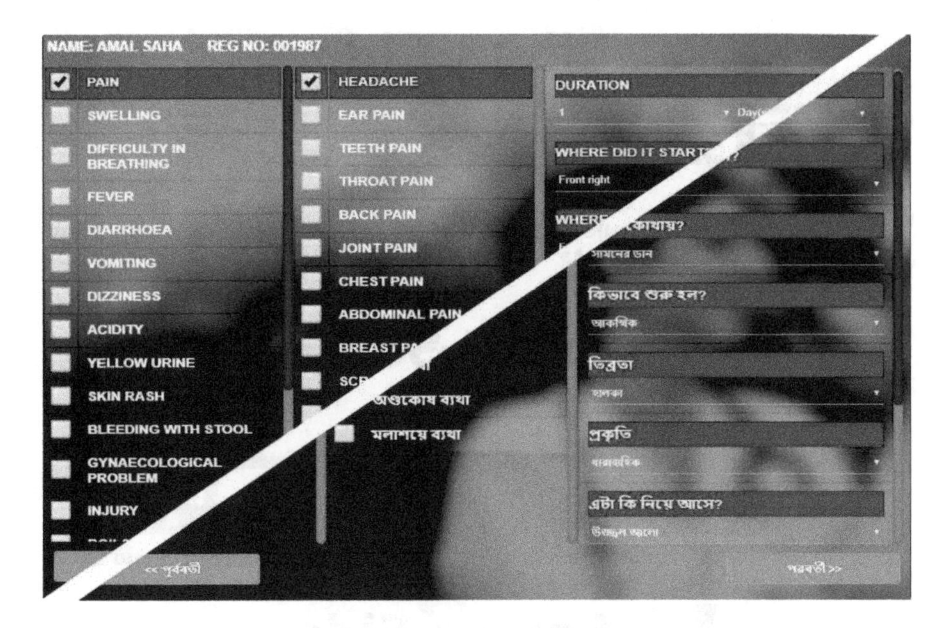

FIGURE 8.6 Symptom collection screen English/Bengali.

A multidimensional hash table is used for storing and showing the knowledge base. This table maintains the key-value pairs for storing symptoms, sub-symptoms, and their related questions and answers.

- **Multilingual Interface:** The application is developed for rural and remote areas for providing primary healthcare services to the rural people. For that reason, application interfaces must support the local languages. Thus, multiple language interfaces can be plugged into the application as per the requirement in a region. Currently, only Bengali and English language support is set up. Figure 8.6 shows the symptom collection screen in English and in Bengali.

  To display and switch between multiple languages, a JSON table is maintained to show the multilingual texts. The JSON table has the collection of different languages that the application supports. Each such language maintains a dictionary that maps a fixed set of key-values corresponding to that language. The set of keys is fixed for all languages. There is a method which takes "language" and "key" as an argument and returns the text phrase for the required language from the JSON table.
- **SMS-based transmission:** The major constraint in the rural and remote areas is the non-availability of reliable internet connection with higher bandwidth. Therefore, in our system, whenever there is a discontinuity in internet connection, the application sends the important data (particularly symptoms and vitals of the patient) to the cloud using SMS instead of the internet data. The data are converted to a JSON string, and the string is compressed and sent through SMS to an SMS server. The SMS server sends

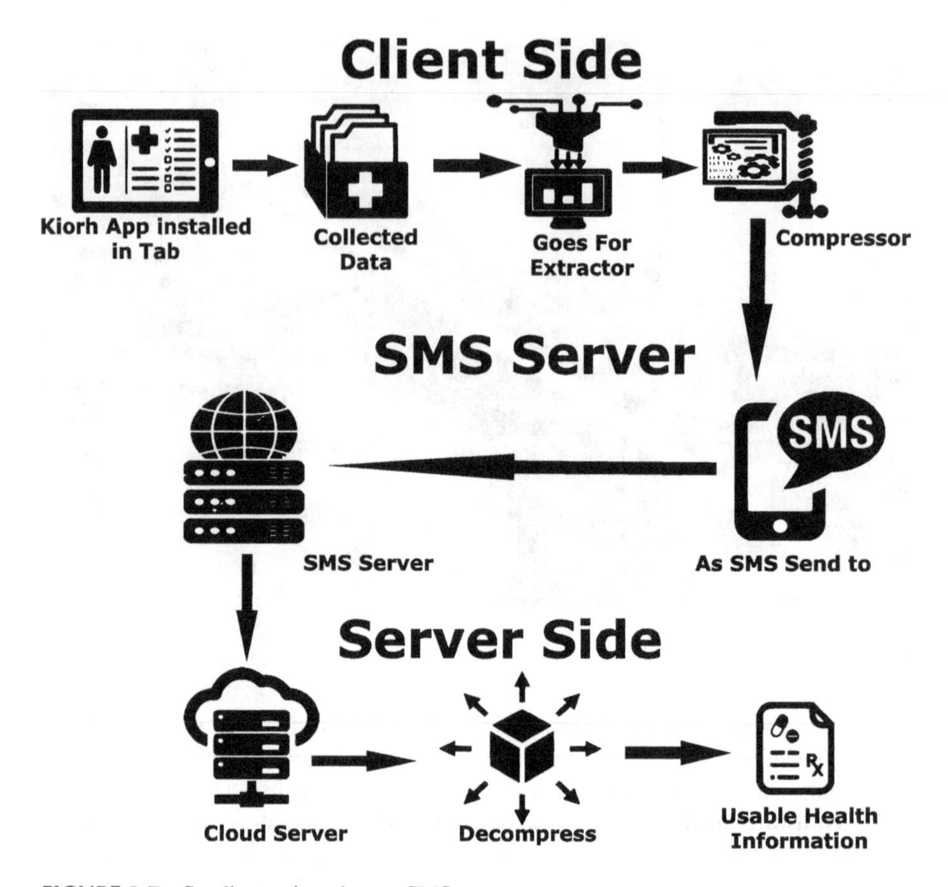

FIGURE 8.7 Sending patient data as SMS.

the data to the application's cloud server, the cloud server decompresses the data, changes it to the original JSON format, and stores in the server for further use. The process is shown in Figure 8.7.

- **Connecting with doctors in real-time:** On the other side of the application, there is a pool of urban doctors. Based on a predetermined schedule, the doctor can login to the system and gain access to the vitals of the patient in real time when a health assistant connects the patient with the doctor. The doctor can view all related and submitted complaints and prescriptions and also old records (if any), such as prescription, from the cloud storage.
- **Prescription generation:** The doctor can prescribe medicines by filling through the web interface (Figure 8.8). The doctor needs to fill some of the fields, like type, name, dose, unit, termination, count. etc., as they are mandatory. The doctor can suggest the mode of medication (clocked-based or event-based, e.g. ("after" (when) and "dinner"' (context)) from the web interface. The application creates prescriptions in multiple languages for accessibility of the users.

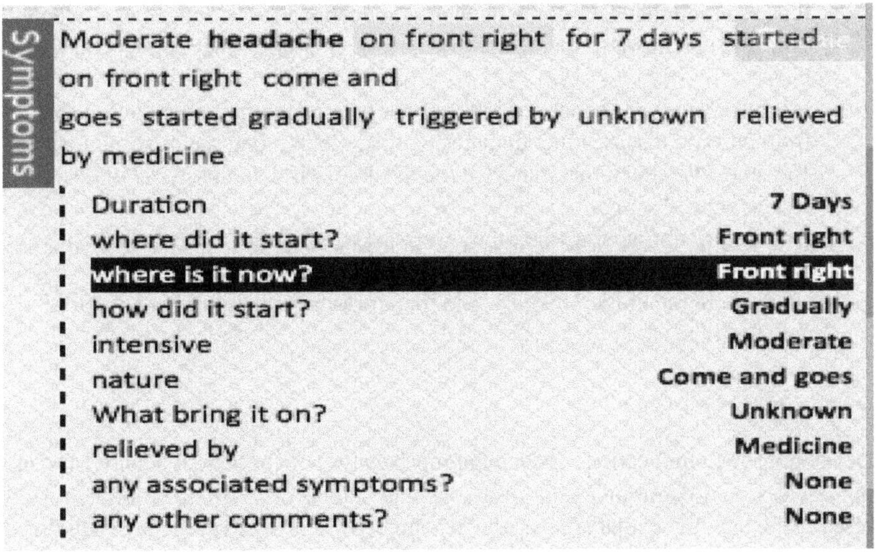

Symptoms

Moderate **headache** on front right for 7 days started on front right come and goes started gradually triggered by unknown relieved by medicine

Duration	7 Days
where did it start?	Front right
where is it now?	Front right
how did it start?	Gradually
intensive	Moderate
nature	Come and goes
What bring it on?	Unknown
relieved by	Medicine
any associated symptoms?	None
any other comments?	None

FIGURE 8.8 Doctor's view of the patient's complaint.

8.6 IMPACT ON THE SOCIETY

We have deployed the KiORH application in the villages in the state of West Bengal in India. Two health kiosks have been set up and the applications have been used for years. Later, a survey has been conducted by the Department of Computer Science and Engineering and School of Mobile Computing and Communication, Jadavpur University, to understand the impact on the villagers. The survey has been conducted among the villagers who visited the kiosk at least once. One hundred random samples have been taken from the registered patient list, and face-to-face interviews have been conducted using pre-tested, structured, and close-ended questionnaires in vernacular language. Another one hundred samples have been picked up from the electoral roll of the village to understand the perception of the villagers who never visited the kiosk. The findings from this survey are summarized below.

- According to the survey, 41% of kiosk goers are in the 35–50 year age group, 24% are in the 24–30 year age group, and 27% of the population is of more than 50 years of age. Only 8% falls in below the 20 years age group. The results show that the persons in the age group 35–50 have utilized the facility more than the persons in other age groups.
- As per gender distribution, the male patient percentage is slightly higher than female patients.
- It is noticed that among the kiosk goers, 65% are graduates and 29% of the kiosk goers are non-literate. This may be the result of the awareness and hesitation-free attitude of the educated section toward such types of online healthcare concept.

- It is observed that 72% of the patients come from a low-income group. Rich villagers can move to the nearby towns and, therefore, they rarely visit the kiosk.
- It is also observed that almost 59% of the patients going to the kiosk suffer from chronic diseases like diabetes, colitis, etc. Among the other patients, 16% and 14% stated that they visit the kiosk for viral and bacterial diseases, respectively.
- Among the non-goers, when they were asked why they have never visited the kiosk, almost 44% cited the reason that doctors or nursing staff are not available in the kiosk. Only 3% said that the kiosk is helpful, but not affordable for them.

8.7 CONCLUSION

Developing IoT applications is challenging because of the heterogeneity of sensing devices, and communication protocols and standards. Virtualization is a well-researched technique that allows the resource owners to expose the virtualized resources in a cloud environment to be accessed on a demand basis. The resources can also be shared by multiple applications transparently. In this chapter, we describe an architecture which hides the underlying hardware details and allows the system to expose the virtualized sensing devices at the IaaS layer of the sensor-cloud framework. We have implemented a healthcare delivery application on top of our sensor-cloud framework and deployed the application in rural areas of West Bengal, India. We also observed the impact of such a sensor-cloud based application for remote healthcare delivery in order to assess its usefulness.

REFERENCES

1. P. Albertos and G. C. Goodwin, Virtual sensors for control applications. *Annual Reviews in Control*, vol. 26, no. 1, pp. 101–112, 2002.
2. P. Ciciriello, L. Mottola and G. P. Picco. Building virtual sensors and actuators over logical neighborhoods. In Proceedings of the international workshop on Middleware for sensor networks (MidSens '06). Association for Computing Machinery, New York, NY, pp. 19–24, 2006.
3. L. Mottola and G. P. Picco. Logical neighborhoods: a programming abstraction for wireless sensor networks. In Proceedings of the Second IEEE international conference on Distributed Computing in Sensor Systems (DCOSS'06). Springer-Verlag, Berlin, Heidelberg, pp. 150–168, 2006.
4. N. Raveendranathan, S. Galzarano, V. Loseu, R. Gravina, R. Giannantonio, M. Sgroi, R. Jafari and G. Fortino. From modeling to implementation of virtual sensors in body sensor networks. *IEEE Sensors Journal* 12, no. 3, pp. 583–593, 2011.
5. G. Buondonno and A. De Luca, Combining real and virtual sensors for measuring interaction forces and moments acting on a robot. In 2016 IEEE/RSJ International Conference on Intelligent Robots and Systems (IROS), Daejeon, pp. 794–800, 2016.
6. J. Kullaa, Structural health monitoring using a large sensor network and Bayesian virtual sensors. In Proceedings of the 1st International Conference on Advances in Signal Processing and Artificial Intelligence, IFSA Publishing, 2019.

7. D. Masti, D. Bernardini and A. Bemporad. Learning virtual sensors for estimating the scheduling signal of parameter-varying systems. In 2019 27th Mediterranean Conference on Control and Automation (MED), IEEE, pp. 232–237, 2019.
8. C. Sarkar, V. S. Rao, R. Venkatesha Prasad, S. N. Das, S. Misra and A. Vasilakos, "VSF: An energy-efficient sensing framework using virtual sensors. IEEE Sensors Journal, vol. 16, no. 12, pp. 5046–5059, June 15, 2016.
9. M. Lemos, C. de Carvalho, D. Lopes, R. Rabelo and R. H. Filho, Reducing energy consumption in provisioning of virtual sensors by similarity of heterogenous sensors. In 2017 IEEE 31st International Conference on Advanced Information Networking and Applications (AINA), Taipei, pp. 415–422, 2017.
10. T. Ojha, S. Misra, N. S. Raghuwanshi and H. Poddar, "DVSP: Dynamic virtual sensor provisioning in sensor–cloud-based internet of things. *IEEE Internet of Things Journal*, vol. 6, no. 3, pp. 5265–5272, June 2019.
11. M. Elhoseny, G. Ramírez-González, O. M. Abu-Elnasr, S. A. Shawkat, N. Arunkumar and A. Farouk, Secure medical data transmission model for IoT-based healthcare systems. *IEEE Access*, vol. 6, pp. 20596–20608, 2018.
12. T. Wu, F. Wu, J. Redouté and M. R. Yuce, An autonomous wireless body area network implementation towards IoT connected healthcare applications. *IEEE Access*, vol. 5, pp. 11413–11422, 2017.
13. R. Mahmud, F. L. Koch and R. Buyya. Cloud-fog interoperability in IoT-enabled healthcare solutions. In Proceedings of the 19th International Conference on Distributed Computing and Networking (ICDCN '18). Association for Computing Machinery, New York, NY, pp. 1–10, 2018. DOI: 10.1145/3154273.3154347
14. M. Harini, K. Bhairavi, R. Gopicharan, K. Ganapathy and V. Vaidehi, Virtualization of healthcare sensors in cloud. In 2013 International Conference on Recent Trends in Information Technology (ICRTIT), Chennai, pp. 663–667, 2013.
15. S. Bose and N. Mukherjee, Sensiaas: A sensor-cloud infrastructure with sensor virtualization. In 2016 IEEE 3rd International Conference on Cyber Security and Cloud Computing (CSCloud), IEEE, pp. 232–239, 2016.

9 Opportunities and Challenges of IoT-Based Smart City Models toward Reducing Environmental Pollution

Ajanta Das, Shubham Prasad, and Sameya Ashraf
Amity Institute of Information Technology,
Amity University, Kolkata

CONTENTS

9.1 Introduction 153
9.2 Related Work 155
9.3 Features of Smart City 156
9.4 Opportunities of Existing Smart City Models 158
9.4.1 China 158
9.4.2 Spain 158
9.4.2.1 Transport 159
9.4.2.2 Energy 160
9.4.2.3 Waste Management 160
9.4.2.4 Environment 160
9.4.3 Brazil 160
9.4.4 England 161
9.4.5 Austria 161
9.4.6 India 162
9.5 Comparison of Challenges and Action Plan of Smart City Models 163
9.6 Analysis of Environmental Pollution Data for Kolkata, Smart City 163
9.7 Conclusion 168
References 169

9.1 INTRODUCTION

A smart city is a concept of a digitized city, which is connected through a distributed wireless sensor network, to provide ease of life to its citizens with minimized effort. Nowadays, with the advent of technology, the digitized world becomes ubiquitous to the people, and each citizen who is willing to improve their lifestyle with the help of internet connection and mobile app. This can be achieved by building a smart system, which will retrieve data from multiple distributed sources, integrating of these large heterogeneous datasets properly, applying adequate transformation rules, extracting information from various observations, proposing new methods for data mining to deal with Big Data, and finally, generating various predictive models. Hence, smart cities are also regarded as a catalyst for the improved and enhanced utilization of various technologies, like sensor network, cloud computing, Big Data intelligence, etc., which will enrich the life of the citizens.

The challenge of urbanization is a big motivator for the development of smart cities. Cities are engines of growth for the economy of every nation. With the rapid growth of urbanization, the increase of a higher percentage of the population is also expected in urban areas [1]. Comprehensive development of physical, institutional, social, and economic infrastructure is required to address the challenge. As a whole, the quality of city life should be improved. The main aim of a smart city is "housing for all." More than 70% of the Indian population would be living in cities by 2050. Due to the increasing urbanization, a better standard of living is extremely needed, which can be achieved by a sustainable model of housing, present in the smart city.

An unsustainable pollution level is created by the industrialization of the developing world [2]. This environmental pollution includes air, noise, water, and soil, etc. The aim of the controlling environment pollution must reduce the carbon emission targeting air pollution. Survey reports that due to air pollution, China and India are having the maximum rate of death. Moreover, at least millions of people are victimized throughout the world due to the harmful effect of air pollution.

Waste pollution is also another alarming issue in today's urban life, which is mainly created by the unprecedented growth of population. Poor traffic management systems cause various problems like unsafe movement by the pedestrians, which causes increasing accident levels for the special group of people like old, children, and disabled [8]. Insufficient energy management greatly affects air, water, and land quality, which ultimately impacts civilization. Energy should be managed properly as the living beings are the only sufferers for climate change, biodiversity loss, desertification, etc., due to global warming.

Sources of air pollution are identified as lack of controlling of carbon emission, thereby huge emission of particulate matter and toxic gases. Instead of reducing or controlling the identified sources, it increases day by day with more urbanization and industrialization [3]. In order to live in a healthy environment, the smart city mission needs to be attempted and achieved.

The objective of the chapter is to present a challenge analysis of a variety of smart city models in the world. It also presents the survey of 24×7 monitoring data of the environment, specifically air quality index (AQI), accessed from [4] and [5], for various regions in Kolkata.

Organization of the chapter is as follows. Related work presented in Section 9.2 basically describes a few features of the smart city, which are already implemented or planned to be implemented across various cities in the world. Section 9.3 presents the features of the smart city, while Section 9.4 discusses the various smart city models. A comparative study of the challenges of all of these smart city models, with proposed action plan for solutions of these challenges, are presented in Section 9.5.

Section 9.6 presets the analysis of air quality monitoring data for ten different regions in Kolkata. Section 9.7 concludes the paper with proposed action plans and expected outcome of the review, analysis, and the action plans.

9.2 RELATED WORK

This section presents a review of various worldwide smart city research studies for different countries. Thereafter it presents the plan for Smart City Mission in India also.

All the traffic solutions provided by KORE gives global wireless connectivity through cellular networks, and delivers around 100% coverage through a single source. KORE delivers the traffic solutions, which can be installed with a communication device, and configured to communicate across wireless networks, so that agencies can take a decision easily as to whether to use the device along with its monitoring capacity regarding the nearby area of the device. So traffic data can be used as the basis of decision making about the city traffic safety condition so that proper precautionary measures can be taken automatically if required [6, 7].

IBM brought in the concept of "Smarter Planet" in China in the year 2009. Since then, the government of China has approved 193 smart city projects across the country [8, 9]. In China, the development of smart cities is viewed as a strategy that must be adopted for the acceleration of urbanization, industrialization, etc. The success of the smart city projects in China is due to the collaboration of the various IT companies with the government of China.

In Spain, a smart city is defined as a very broad concept whose basic capability lies in providing a wide range of services to the various clients. The important features considered within a smart city project in Spain are: energy-efficient constructions, smart communication grids, use of renewable energy, efficient management of water, mobility sustenance, and electric vehicles. The government of Spain has made heavy investments toward renewable energy development, increasing energy efficiency, and reduction of emission of greenhouse gases [10].

The main objective of the Brazil government in undertaking the smart city project in Brazil, was to bring itself to the forefront in terms of urbanization in the world. The most famous city of Brazil, Rio de Janeiro, had to face various challenges in security, unpredictable climatic conditions, etc., before the smart city concept was introduced there. So, Brazil attempted to overcome these challenges by making use of smart technologies and thereafter converting Rio de Janeiro into a smart city [11].

In Austria, it is believed that smart cities are the remedy for dealing with various issues in energy and unpredictable climatic conditions affecting the country [12]. Also, the tremendous pressure provided by economic and technological modernization across the world leads to the initiation of smart city projects in Austria. The target areas for the smart city project undertaken in Austria are: promotion of cutting-edge technologies of Europe, for ensuring sustainable supply of energy, for reduction of carbon emissions.

India's target in taking up the smart city project [13] is to achieve the goal of providing "Smarter Solutions for a Better Tomorrow." The population of the country has undergone a tremendous growth in recent years, and the smart city initiative has therefore been taken up to cater to the needs of improvement of the quality of life for the present as well as future generations. The government has assigned 70.6 billion dollars (USD 1.2 billion dollars) to the "Smart Cities" project in India in the budget of the year 2014–2015. India has picked out one hundred cities from among the various states and union territories of the country for implementing the smart city project.

9.3 FEATURES OF SMART CITY

Smart City means that a city is connected with the digital world; i.e., the facilities can be searched or utilized through proper internet connection. In a nutshell, it is a concept which provides its citizens improved quality of life with increasing operational efficiencies. However, the overall cost for this massive implementation should minimize as well, as it will provide for more business and minimizing adverse environmental impact. Figure 9.1 presents the features for any proposed smart city. This figure shows seven different wings, which need to be improved parallel to make the city smart. First of all, the *infrastructure* needs to be the first one to be achieved. If the proper infrastructure is built, then only *ICT* and *political participation* are accessible. *Environmental* and *social* are two most important wings, where citizens can live with better health. The people can think for innovation or do perform intellectually better if they deal with a proper healthy life and can have leisure time as well. They need to participate in the cultural program also to think better. Hence, last but not the least, are two other wings: *cultural and leisure* and *intellectual and innovation*.

This section highlights the basic features of the smart city and how the features are implemented in real life. Actually, the features will be implemented using details

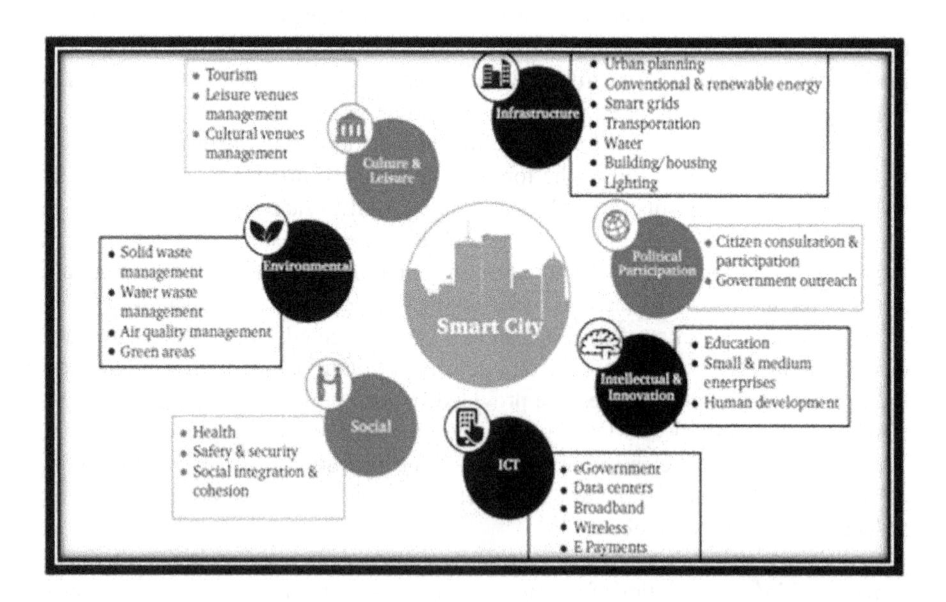

FIGURE 9.1 Features of smart city.

and some recent advanced technology. This section correlates the specific feature with relevant technologies to be used.

The basic features of the smart city are described below.

Smart parking: This system will alert a driver about a free parking spot available in nearby regions. It will result in reduction of carbon dioxide emission by about 400 tons in a city, simply by the reduction of the average time required to search for a parking spot for a vehicle by approximately 7 minutes instead of 10–12 minutes.

Smart healthcare: Online applications are available which take care of the patients in the absence of doctors, but doctors will be consulted through mobile if the patient's condition deteriorates with time. Alerts will be sent to the doctors if certain parameter changes, like blood pressure or sugar rise, occur. Recent day's telemedicine is also helpful in some other situations.

Doctors will be providing suggestions online if they are not present in the emergency situation. Various alertness programs can be shown online to the citizen to increase the awareness.

Smart traffic management: An automated road monitoring system will provide information to the driver regarding the available best suitable route at any given time. Also, the traffic lights will be managed automatically to make the congestion minimum at certain peak hours.

Smart energy management: This feature can be divided into two set of tasks: *smart grid* and *smart urban lighting*. The electricity requirement for necessity and regular life will be provided by the *smart grid,* and the light intensity may be adjusted based on the presence of the people in the surrounding through *smart urban lighting*.

Smart waste management: Smart waste management includes smart bin, and conversion of waste to energy or fuel and recycling. Smart bins can be used, which will inform automatically when it should be emptied. Usages of papers need to be increased instead of plastics to recycle them and conversion to energy.

Smart containers, including a fleet management system, can be efficiently used for the waste collection, and the route may be customized as and when necessary based on the requirement of the situation.

Smart city maintenance: Maintenance for any damages in the urban elements can be faster through notification by the citizens to the City Council through their smartphones.

Smart taxi: A mobile app is very useful to access the facility of smart taxi. The localization system will notify the closest one without the need of human interaction. Some mobile applications are already available for providing this kind of service, like Ola, Uber, etc.

Smart ticketing system: Digital signage can be an option for the customers to save the time so that they can buy the ticket for travel and entertainment via the billboard instead of visiting the actual place.

Customers can buy and verify tickets using their smartphone, which drastically reduces the waiting time for the passengers in the queue.

Intelligent transport system: Public transport vehicles need to be connected with each other, for receiving coordinated information from them in real time.

Intelligent water management: The quality of water needs to be monitored in regular intervals to improve the quality of life of the citizen. The smart meter and water quality monitoring system will collect water samples from various parts of the city, analyzes it, and send a report to the government authorities.

e-Governance: Citizen services should be improved through e-Governance, which will provide the proper public information for a smart city and the dealings of quality life in a healthy environment.

9.4 OPPORTUNITIES OF EXISTING SMART CITY MODELS

A few cities in the world have taken the initiative to implement identified or targeted features of a smart city in their models. This section presents all these various smart city models, mentioning the identified features and objectives behind all those features, and strategies for implementation or initiatives in the following.

9.4.1 China

The former president of China, Hu Jintao, made a commitment regarding decrease of carbon emissions by 40–45% in 2002 as compared to the emission rate in 2005 [6, 7]. The trigger behind the smart city initiatives undertaken by China is the concept of "Smarter Planet," which was introduced in the year 2009 by IBM. The objective of the above initiative was to carry out 193 pilot projects, which got approved by 2013, targeted toward the development of smart cities in China.

In China, the development of smart cities is viewed as a fruitful strategy that must be adopted for the acceleration of industrialization, urbanization, etc. Smart cities are also regarded as a catalyst for the improved and enhanced utilization of various technologies like cloud computing, Big Data, etc. The target area for taking up smart city projects in China therefore is:

- Improvement of urbanization technologies, and
- Promotion of the industrial transformation taking place across the country.

This research inspects the major actors involved in smart city projects in China and their contribution, presented in Figure 9.2. Construction of smart cities requires development across four layers: *sensor layer, network layer, platform layer,* and *application layer.*

The major actors involved in the development of smart cities are as follows:

- *Government*: Nearly 180 cities have joined the initiative "to become smart" and the government has made an investment of almost 500 billion RMB.
- *Society*: The public sector units cooperate in the smart city projects in an indirect way. For example, the society cooperates with the government and the market by buying their products and/or services [8].

9.4.2 Spain

The objective is the development of various smart cities in Spain.

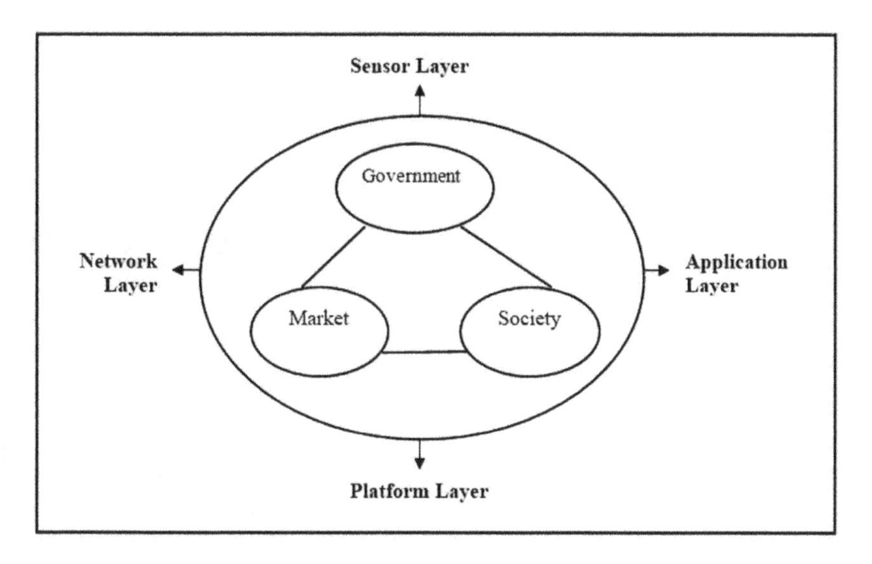

FIGURE 9.2 A conceptual framework on actors of each layer [8].

Spain has participated in the smart city project with a view to improve the following features of a city [10]:

- Energy-efficient constructions
- Smart communication grids
- Use of renewable energy
- Efficient management of water
- Mobility sustenance
- Electric vehicles

In Spain, heavy investments have been made toward renewable energy development, increasing energy efficiency, and reduction of emission of greenhouse gases.

A smart city example in Spain is Malaga, as presented in Figure 9.3. In Malaga, many smart technologies have been applied, for example:

- Lighting improvements in public places
- Systems for efficient management of energy, etc.

EU-funded Smart City Project for the city of Santander has already been completed recently with major initiation from the University of Santander and urban service providers [6, 7].

This can be told as the largest strategic innovation in the continent.

Below are the target regions which have been considered in the above plan.

9.4.2.1 Transport

For better traffic management, sub-surface geomagnetic sensors have been used in parking bays, and the data thus collected helps to avail car parking space efficiently.

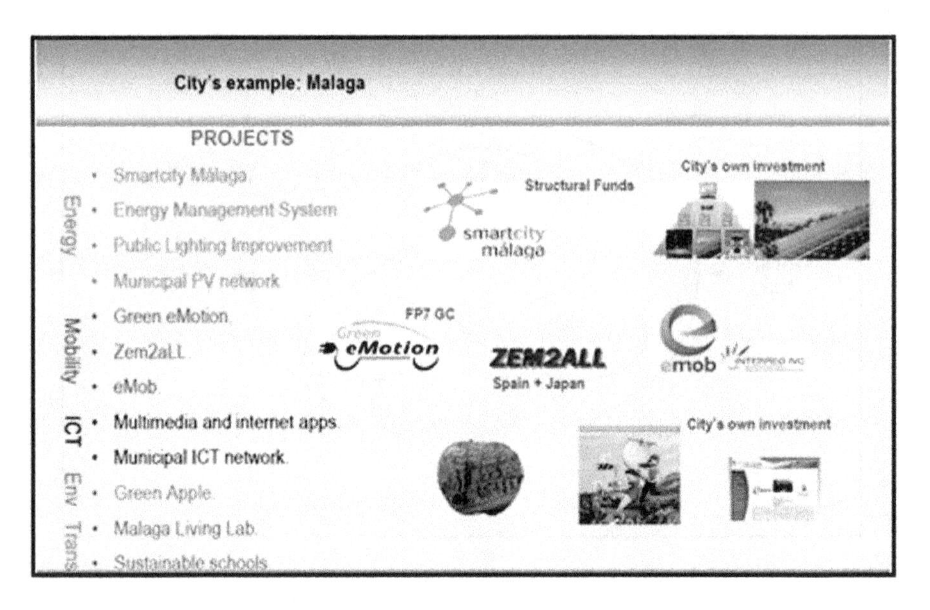

FIGURE 9.3 Smart city Malaga in Spain [10].

Similarly, traffic jams and accidents can be tracked in real time from the data retrieved from sensors, etc., for help in urban planning.

9.4.2.2 Energy

Various techniques for the smart consumption of energy have been developed, which will increase the light intensity based on the number of surrounding people.

9.4.2.3 Waste Management

Smart bins fitted with sensors are planning to be implemented, which will alert the local authorities when the bins need to be emptied.

9.4.2.4 Environment

Various types of equipment are placed across the city to measure the air quality and noise levels automatically. Based on this measurement, various rules can be applied to meet the city guidelines.

9.4.3 Brazil

The objective is the development of various smart cities in Brazil to bring itself to the forefront in terms of urbanization in the world, and to overcome the various challenges in security, unpredictable climatic conditions, etc., faced by the most famous city of Brazil, Rio de Janeiro, by making use of smart technologies and thereafter converting into a smart city.

In 2010, the municipality government of Rio de Janeiro opened a functional unit named *Centro de Operações* (a Portuguese term meaning Centre of Operations).

This center included offices meant for the surveillance of traffic, forecasting of weather conditions, etc. This was done by collaboration with companies like IBM, Oi, Cisco, and Samsung.

In 2012, during the rainy season, prompt actions undertaken with the help of the Centre of Operations helped to ensure that no casualties were taking place due to floods or landslides. Also, reports from the same year show that there had been a 25% reduction in time taken for responding to accidents caused due to the traffic.

Therefore, undertaking this smart city project helped Rio de Janeiro to be announced as "the leading smart city in the world" at the Smart City Expo World Congress in 2013 [11].

The Sao Paulo City of Brazil plans to implement intelligent monitoring of lighting systems making use of Ilumatic lamps, which have been integrated with CyLux® technology for cost savings [6, 7].

9.4.4 England

Smart Transportation system is a very important feature of a smart city where the safety of the pedestrians is a major aspect. London is the first city that plans to implement crosswalks integrated with sensors, and cameras to make the streets smart, so that they can be crossed easily and safely.

Mostly, the traffic signal works on an interval basis, i.e., it displays a red light for stopping the traffic for a particular side for t seconds, and then it remains green for t seconds for allowing traffic in the same side also. However, the smart transport system may extends green pedestrians through automatically adjusting the signal timing, so that many people can cross the road with less waiting time. As an example, if a large number of people are waiting, the time for green light for pedestrians should also be increased accordingly.

Apart from this, the Westminster Council has started planning to implement smart parking sensors across its entire road network. The sensors will automatically detect whether the parking bay is vacant. The drivers can use the real-time map of parking spaces to find the vacant space. The real-time map will come automatically by installing the application. The improved system should be able to decrease both congestion and carbon emissions successfully. The authorities have the plan to implement this project across various cities if implemented successfully in Westminster [6, 7].

9.4.5 Austria

Vienna as a proposed smart city in Austria is using strategic energy and technology plans, as presented in Figure 9.4.

The target areas for the smart city project undertaken in Austria are [8]:

- Promotion of cutting-edge technologies of Europe
- For ensuring sustainable supply of energy
- For reduction of carbon emissions

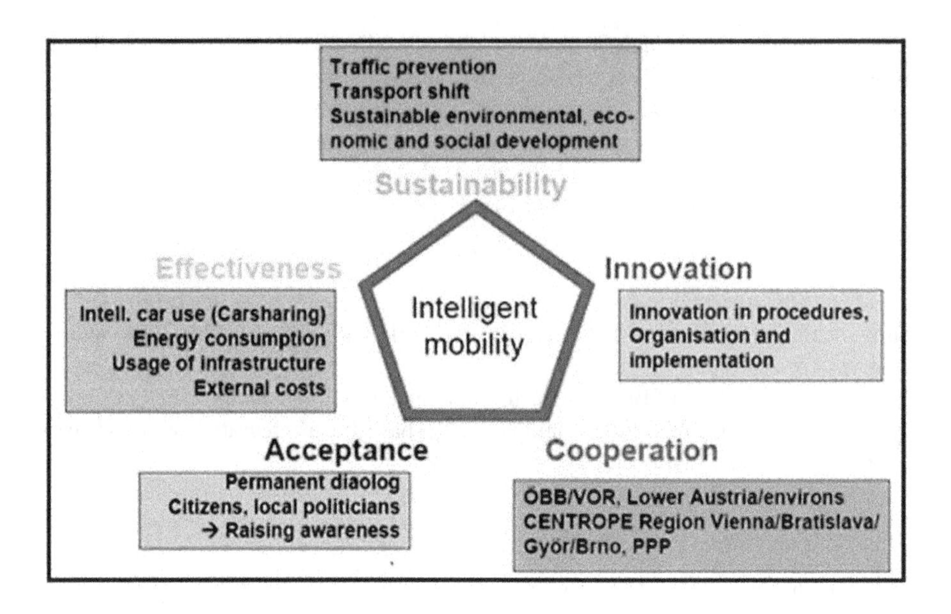

FIGURE 9.4 Vienna's transport master plan [8].

The various programs undertaken to make Vienna "smart" are:

- *Urban energy efficiency program:* As the name implies, it aims for the efficient and sustainable use and management of available energy reserves.
- *Eco buy Vienna:* Five billion euros have been invested in various products as well as services to improve their quality based on some ecological criterias. This has resulted in carbon saving of about 135 tons of carbon dioxide between the period of 2005 and 2008.
- *Vienna's transport master plan:* This includes strategies like introduction of green buses, recuperation of energy, etc.

9.4.6 India

In order to achieve the goal of providing "Smarter Solutions for a Better Tomorrow," the citizens of the country have undergone a tremendous growth in recent years, and the smart city initiative has therefore been taken up to cater to the needs of improvement of the quality of life for the present as well as future generations.

The smart city initiatives have been focused on the following areas [13]:

- Management of sufficient supply of water.
- Effective distribution of electric supply.
- Employing proper techniques in waste management.
- Maintaining good transport facilities, etc.
- Reducing of Air Pollution Level.

The government has assigned 70.6 billion dollars (USD 1.2 billion dollars) to the "Smart Cities" project in India in the budget of the year 2014–2015. India has picked out a hundred cities from among the various states and union territories of the country for implementing the smart city project. Some major cities in India thus identified are: Lucknow, Guwahati, New Town Kolkata, Ranchi, Jalandhar, Jaipur, Gandhinagar, etc.

Surat has been selected to be developed as India's first smart city under the initiative of Microsoft Corporation. Citizens will be updated in live and real-time data through the blending of cutting-edge technologies like cloud computing, mobile computing, data analytics, and social networking provided by Microsoft. There is also a plan of developing an automatic system, which can protect the city from the effect of natural disaster to some extent [6, 7].

9.5 COMPARISON OF CHALLENGES AND ACTION PLAN OF SMART CITY MODELS

This section briefly compares the challenges based on the discussion in the above section. With reference to the availability of data, various countries have identified the challenges for their corresponding smart city mission. A choice of challenges includes carbon emission rates, energy management, use of renewable energy, waste management, smart parking, transportation management, and environmental or air pollution, etc. Simultaneously, the action plans are also proposed to overcome the challenges safely and ubiquitously.

Table 9.1 describes various smart city features, which are implemented across the globe. This table represents various challenges with its action initiator and proposed action plan.

The next section presents a survey of environmental pollution, specifically air pollution or AQI data, to achieve in the smart city mission of Kolkata, India [14].

9.6 ANALYSIS OF ENVIRONMENTAL POLLUTION DATA FOR KOLKATA, SMART CITY

In order to reduce air pollution, intense monitoring needs to be conducted to measure the air pollution generated from various sources. Moreover, it is necessary to assess or predict air quality, so it will monitor and assess air pollution keeping Continuous Air Quality Monitoring Station (CAQMS) at various centers. The chapter presents a thorough survey of the overall AQI of various regions or zones, where the air quality is monitored through the West Bengal Pollution Control Board. Contribution to environmental air pollution is gradually increasing day by day with more industries, vehicles, and housing projects. Specifically, with the increase of housing or realtor projects, the numbers of habitants or populations are actually contributing to environmental pollution through more vehicles, deforestation, and traffic congestion in that particular area.

TABLE 9.1
Challenges for Worldwide Smart Cities with Proposed Action Plan

Country (City)	Challenge	Action Executor	Proposed Action Plan
China (Shenzhen and Ningbo)	Higher carbon emission rate; Inefficient energy management	The Ministry of Housing, Urban and Rural Development (MOHURD) of China	Plan to develop a low carbon city; Making clean energy
Spain (Malaga)	Energy-efficient constructions, smart communication grids, use of renewable energy	Spanish Government	Heavy investments have been made toward renewable energy development, increasing energy efficiency, and reduction of emission of greenhouse gases
Spain (Santander)	Smart parking, efficient energy, waste management	University of Santander and urban service providers	Installing a network of 12,000 sensors of different types on buildings, buses, and a range of infrastructure to collect information about the parking space and share this information to the drivers through smartphone
Brazil (Rio de Janeiro)	Security, unpredictable climatic conditions	Brazil Government	The municipality government of Rio de Janeiro, opened a functional unit named Centro de Operações (a Portuguese term meaning Centre of Operations), which included offices meant for the surveillance of traffic, forecasting of weather conditions, etc.
Brazil (São Paulo)	Inefficient lighting	Cyan Holdings plc and Ilumatic S/A	Plan to implement intelligent lighting control monitoring using Ilumatic lamps integrated with CyLux® technology for cost savings and low carbon emissions.
England (London)	Transportation problem: unsafe pavement	Government authorities	London is the first city that plans to implement crosswalks integrated with sensors, and cameras to make the streets smart so that they can be crossed easily and safely.
Austria (Vienna)	Promotion of cutting-edge technologies of Europe and reduction of carbon emissions	Austrian government	The various programs undertaken to make Vienna "smart" are: • Urban energy efficiency program • Eco buy Vienna • Vienna's transport master plan
India (Surat)	Poor traffic management, unavailability of parking spaces	Initiative of Microsoft Corporation	Making "Surat" a smart city to provide information from real-time data to all the citizens of the city

The major causes for air pollution are listed in the following:

- Using poor quality fuel in vehicles, which includes high sulfur, benzene, and olefin.
- Heavyweight old engine petrol-driven vehicles and mixing of kerosene and exhaust mobile gases in the engine of three-wheelers causes high emission of pollutants.
- A poorly maintained and damaged old vehicle supplemented by erratic traffic behavior, which leads to congestion and emits a huge proportion of automobile pollutants.
- With the change of season and weather, concentration of air quality varies, and it is more severe during winter.
- More industrialization and sanction of more housing projects increases emission of carbon dioxide level and decreases of oxygen.

AQI is monitored for last six months, from September 2019 to February 2020. The source of the monitoring data is collected from the authorized controlling body, West Bengal Pollution Control Board [5]. This chapter identifies three automatic stations and seven manual stations scattering throughout Kolkata city. The automatic station carrying out monitoring 24×7 or every instance changes of the data may be reflected. In contrast, the manual station publishes monitoring data for a few particular days for the specific month. Reasons behind the monitoring scheme may be mentioned like this, that automatic stations are located in high-road congested, busy, eventful, and highly populated places. Whereas the manual stations are located comparatively less crowded and almost mostly having no road congestion places. So, in the case of automatic stations, average AQI data is presented. Moreover, the highest and lowest AQI are also presented for the reference. Average AQI is calculated for each month, based on the collection of data on the particular day for that month for each station. More specifically, AQI data is collected from National Air Quality Index [4] in every four hours starting from 1 A.M. to 1 P.M. for each *Monday* during the month of *February*, each *Tuesday* during the month of *October*, each *Wednesday* during the month of *January*, each *Thursday* during the month of *September*, each *Friday* during the month of *December,* and each *Saturday* during the month of *November.* The AQI data for these last six months of the locations, Ballygaunge, Jadavpur, and Rabindrabharati, are represented in Table 9.2.

Similarly, in the case of manual stations, the AQI is monitored for the last six months, from September 2019 to February 2020. The AQI data are collected from the West Bengal Pollution Control Board under the Manual stations [5]. In the case of a Manual station, AQI data are presented once for a whole day, and the air pollution monitoring is carried out specifically for a few mentioned days in each month. It implies that air quality is not monitored for all the days in each month. Here, the highest and lowest AQI data of seven locations, like Baruipur, Minto Park, Moulali, Paribesh Bhawan, Rajarhat, Salt Lake, and Tollygunge, are presented in Table 9.3.

Observations: The lowest data are observed during the night between 10 P.M. and 1 A.M. or early in the morning during 4 A.M. to 5 A.M., except during the period of the Diwali festival. Whereas, in contrast, the highest data are observed

TABLE 9.2
AQI Monitoring Data for Automatic Stations

S. No.	Location	Monitoring Month	High AQI	Low AQI	Avg. AQI
1.	Ballygaunge	February	305	128	237.70
		January	385	189	260.46
		December	258	156	200.69
		November	277	102	185.89
		October	269	96	186.41
		September	34	18	26.91
2.	Jadavpur	February	195	129	164.49
		January	298	81	175.47
		December	149	103	127.16
		November	189	18	125.44
		October	219	22	117.45
		September	33	17	22.79
3.	Rabindrabharati	February	308	238	278.66
		January	399	159	243.99
		December	286	169	223.49
		November	281	43	208.45
		October	246	58	141.21
		September	66	30	44.4

between 6 A.M. and 8 P.M. or 9 P.M. Among all these data of ten stations, the *maximum of lowest AQI* data and *minimum of highest AQI* data are marked red to distinguish them easily. It is also observed, irrespective of the regions and type of the stations, that the minimum of highest AQI measured in the month of September and maximum of lowest AQI measured mostly in the month of February (70%), December (20%), and January (10%).

Six standard categories of AQI exist depending on various levels of monitoring data [4] and [5]. Table 9.4 represents the standard AQI for various groups. Therefore, from the standard AQI index, it is obvious that AQI crossing limit 100 is threatened to the citizen. Hence, this paper [15] highlights that it is always on urgent threat for air pollution. This paper proposes a layered air pollution reduction framework, which presents the directives for social benefit through the expected outcome. This research also suggests some rule base for prohibiting congestion on roads and thereby reducing air pollution. In [16, 17], the suggested and proposed rule base in [15] are implemented, depending on air pollution level, using decision tree and machine learning techniques.

Trees play an important role to reduce air pollution. Due to the increase of industries, public vehicles, and populations in major arterial roads, the plantation of trees on the wide boulevards of the major arterial road becomes a necessity to control air pollution on these regions. According to the news and reports published in [18], there are more than 1,65,000 trees that have been planted in major arterial

TABLE 9.3
AQI Monitoring Data for Manual Stations

S. No.	Location	Monitoring Month	High AQI	Low AQI
1.	Baruipur	February	191	161
		January	230	153
		December	265	137
		November	148	68
		October	103	48
		September	55	40
2.	Minto Park	February	255	210
		January	205	54
		December	201	83
		November	250	36
		October	107	41
		September	72	32
3.	Moulali	February	307	101
		January	306	123
		December	312	133
		November	273	117
		October	315	56
		September	82	44
4.	Paribesh Bhawan	February	124	108
		January	194	49
		December	196	79
		November	158	35
		October	162	42
		September	75	34
5.	Rajarhat	February	186	77
		January	236	105
		December	229	138
		November	228	90
		October	153	44
		September	42	35
6.	Salt Lake	February	129	105
		January	219	72
		December	87	54
		November	115	82
		October	140	39
		September	71	42
7.	Tollygunge	February	115	101
		January	128	100
		December	154	74
		November	132	43
		October	78	47
		September	55	31

TABLE 9.4
Standard AQI and Health Impact [5]

Air Quality Index (AQI)	Remark / Health Impact
0-50	Good / Minimal Impact
51-100	Satisfactory / Minor breathing discomfort to sensitive people
101-200	Moderate / Breathing discomfort to the people with lung, heart disease, children and older adults
201-300	Poor / Breathing discomfort to people on prolonged exposure
301-400	Very Poor / Respiratory illness to the people on prolonged exposure
>400	Severe / Respiratory effects even on healthy people

roads. In this way, the beautification of the city will be also taken care of simultaneously with reduction of air pollution. However, the trees need to fence properly to be protected from the storm; otherwise, the roads will be blocked with sudden fall of these trees on these busy roads. It is also suggested that citizens need to plant a few flexible trees on high roads, such as palm, coconut, cherry blossom, pine, and casuarinas, in front of their houses or residential complex, which will ensure beautification, as well as reduce air pollution all together. So, in order to reduce air pollution, standard practice, suggestive measures, action plans, and rule sets proposed in [15] needs to be maintained properly.

9.7 CONCLUSION

This chapter presents various smart city models stating distinct features and proposed benefits for the citizens. Next, it presents a brief survey for awareness of air pollution level in Kolkata, India. Level of air pollution is always on threat for metro cities like Delhi, Mumbai, Kolkata, and Chennai. The major reasons behind the high level of pollution is identified as ground level ozone and particulate matter, which increases with the increase of number of vehicles, like public transports and tracks. Moreover, with the increased travel, demand has also resulted in rapid growth in the number of motor vehicles in the arterial roads of Kolkata City.

Implementation of the smart city project is expected to have following outcomes:

- Smart management of water, wastes, and solid wastes and sanitation: India has a campaign named as Swachh Bharat Abhiyan, meaning Clean India Mission, whose goal is to make India a cleaner place to live in by 2019.
- Smart tackling of issues facing the environment: To reduce the amount of carbon emissions and also to reduce the fatalities caused by air pollution.
- Utilizing smart techniques in urban planning: To build pedestrian skywalks, cycle tracks, etc., in cities. There is a plan by the government for the construction of two hundred low-cost airport terminals in various towns across the country.
- Development of smart buildings: There has been a registration of around three thousand projects for the construction of green buildings with the

Indian Green Building Council (IGBC), of which around six hundred certified are already in function.

- Development of smart grids: Construction of smart grids will result in the significant reduction of the environmental impacts caused by the current system of electricity supply.
- Generation of cleaner energy: India seeks for the utilization of renewable sources of energy to meet the energy and electricity demands of the country.
- Beautification of cities with planting trees: Kolkata will be clean, more beautiful, and green for healthy life very soon. Concurrently, all the metro cities can adhere these proposed action plans to reduce the overall air pollution level in India.

REFERENCES

1. "Smart Cities – Mission Statement and Guidelines" by the Ministry of Urban Development, Government of India. Retrieved from http://smartcities.gov.in/writereaddata/SmartCityGuidelines.pdf.
2. James O. Wheeler, Yuko Aoyama, and Barney Warf, "Cities in the Telecommunications Age: The Fracturing of Geographies," Routledge, New York, 2000.
3. Mansi Shah, "Waiting for Health Care: A Survey of a Public Hospital in Kolkata." Retrieved from http://ccs.in/internship_papers/2008/Waiting-for-Healthcare-A-survey-of-a-public-hospital-in-Kolkata-Mansi.pdf, accessed on 22nd February 2016.
4. National Air Quality Index. Retrieved https://app.cpcbccr.com/AQI_India/, accessed on 18th March 2020.
5. West Bengal Pollution Control Board, Air Quality Information System. Retrieved from http://emis.wbpcb.gov.in/airquality/filter_for_aqi.jsp, accessed on 18th March 2020.
6. Parul Parmar "Smart Home, Building, and City Machine-to-Machine (M2M) Applications," Mind Commerce, 2014, ReportBuyer, 27 August 2014.
7. Pethuru Raj, and Anupama C. Raman "Intelligent Cities: Enabling Tools and Technology," CRC Press, Boca Raton, FL, pp. 1–491, 2015.
8. Sukla Bhaduri, "Vehicular Growth and Air Quality at Major Traffic Intersection Points in Kolkata City, An Efficient Intervention Strategies," The SIJ Transactions on Advances in Space Research & Earth Exploration (ASREE), vol. 1, no. 1, September–October 2013.
9. Yongling Li, Yanliu Lin, and Stan Geertman. "The development of smart cities in China" in CUPUM, 2015, pp. 1–20.
10. Elena Villalba Mora, "Smart Cities in Spain," Centre for the Development of Industrial Technology Ministry of Science and Innovation, pp. 1–11. https://ec.europa.eu/information_society/activities/sustainable_growth/docs/smart-cities/smart-cities-spain.pdf, 2014.
11. "Smart City Projects in Brazil." Retrieved from https://innovationhouserio.wordpress.com/2015/05/26/smart-city-projects-in-brazil/, accessed on 26 February 2016.
12. Alexandra Vogl, Robert Simburger, and Tina Vienna, "Energy Efficient Measures in the City of Vienna," Municipal Department for Urban Development and Planning, Austria, pp. 1–20, https://archive.corp.at/cdrom2011/papers2011/CORP2011_237.pdf, 2011.
13. "Smarter Solutions for a Better Tomorrow." Retrieved from http://www.smartcitiesindia.com/pdf/Smart_Cities_India_2016_Brochure.pdf/, accessed on 26 February 2016.
14. Jana Spiroska, Md. Asif Rahman, and Saptarshi Pal, "Air Pollution in Kolkata: An Analysis of Current Status and Interrelation between Different Factors," SEEU Review, vol. 8, no. 1, pp. 182–214, February 2013, ISSN (Print) 1409–7001, DOI: 10.2478/v10306-012-0012-7.

15. A. Desarkar, and A. Das, "A smart air pollution analytics framework" in International Conference on ICT for Sustainable Development (ICT4SD2016), Bangkok, 12–13 December 2016, pp. 197–205.
16. A. Desarkar, and A. Das, "Implementing decision tree in air pollution reduction framework" in International Conference on Smart Computing and Informatics (SCI 2017), Visakhapatnam, India, vol. 1, 3–4 March 2017, pp. 105–113.
17. A. Das, and A. Desarkar (2018) "Decision Tree-Based Analytics for Reducing Air Pollution," Journal of Information and Knowledge Management (JIKM), vol. 17, Issue 2, pp. 1–20. ISSN: 0219–6492, June 2018.
18. S. Sengupta, "Earthy plan saves new town trees," in Telegraph Online, March 4, 2019. Retrieved from https://www.telegraphindia.com/states/west-bengal/earthy-plan-saves-new-town-trees/cid/1686167, accessed on 18th March 2020.

10 A Novel QoS-Based Flexible Service Selection and Composition Model for Localizing the Wireless Sensor Nodes Using AHP

Akhilendra Pratap Singh
National Institute of Technology Meghalaya, India

Arun Kumar
National Institute of Technology, Rourkela, India

Om Prakash Vyas and Shirshu Varma
Indian Institute of Information Technology,
Allahabad, India

CONTENTS

10.1 Introduction 172
10.2 Related Work 172
10.3 FSONA-Based Localization Services 175
  10.3.1 Components of the FSONA Framework 175
  10.3.2 Formalization of FSONA Services 176
  10.3.3 Necessities of Service Selection 178
  10.3.4 Service Composition Using BB and Workflow 179
10.4 AHP-Based System Model 179
  10.4.1 Convergence of Service Request 181
  10.4.2 Point-Wise Convergence of Service Request 182
  10.4.3 Almost Sure Convergence of Service Request 182
10.5 Implementation, Experimentation, and Results 182
  10.5.1 Criteria Identification 182
  10.5.2 Creation of Hierarchical Model 183
  10.5.3 Pair-Wise Comparison 183

10.5.4 Identification of Resultant Priority and Decision 184
10.5.5 Results and Discussion 185
10.6 Conclusions 193
References 193

10.1 INTRODUCTION

Service-oriented architecture (SOA) is a paradigm of engineering complex software-based systems that focus on decoupling software functionalities from a specific application [1]. In this age, where internet-based services are the norm rather than the exception, such a service-oriented approach garners rapid application development and deployment [2].

However, selecting services from an existing available pool of services with a black-box approach asks applications to subscribe to the pros and cons of the selected services. On the contrary, collecting detailed implementation-related information and subsequently choosing specific services fails the very purpose of agile development [3]. Besides, it is practically impossible for users to comprehend performance-related implications in choosing a specific service within a reasonable time; particularly in comprehending runtime constraints and identifying significant performance parameters. The scenario becomes even more complicated as the application requirements become more complex [4, 5]. These constraints have paved the way for choosing appropriate services and composing an application workflow in an automated fashion, based on certain criteria.

This chapter presents a multi-criteria-based decision-making plan for appropriate service selection and service composition that accounts for the performance requirements of the application to be developed. The analytical hierarchy process (AHP) enables decision-makers to structure complicated problems as a hierarchical structure. The hierarchical model consists of three logical components, namely: alternatives, criteria, and goals, with their inter-relationships arranged as a top-down hierarchy [6]. The proposed model is used to enumerate the multi-criteria-based decision-making for a localization service in a wireless sensor network (WSN). The authors in [7] used this case for a scenario of available services that are made available using the flexible service-oriented network architecture (FSONA) for WSN. The efficacy of the proposed method is demonstrated in the obtained results.

The remainder of this paper is organized as follows: Section 10.2 presents a brief overview of related research work. In Section 10.3, a detailed discussion of FSONA-based localization services with their relative merits and demerits is presented. The proposed AHP-based multi-criteria decision scheme for service selection and composition is presented in Section 10.4. Section 10.5 presents the experimental evaluation of the proposed scheme with detailed discussion on the obtained results. Section 10.6 concludes the chapter.

10.2 RELATED WORK

This section presents, in brief, the fundamentals as well as the state-of-the-art WSN and AHP-related research work.

WSNs find a wide scope of applications, such as monitoring, natural disaster relief, patient tracking, military targets, and automated warehouses. Deployment of sensor nodes in a large area is a challenging task due to its inherent constraints, such as limited energy, communication cost, computation cost, and accuracy. WSN is designed to sense events and locations from deployed sensor nodes where user intervention is not possible or difficult to deploy [8]. Nodes are deployed in a distributed way and are connected over wireless links via hop-to-hop communication. Localizing a node in a WSN is a fundamental yet very challenging proposition, particularly considering other challenges in a WSN.

Every node in the WSN should know the location of every other node. Such localization serves easy identification of event origin and also node identification and message, control communication, and propagation. This can be achieved using the neighbors, and this process is known as localization. For the localization and setting up of the network, various services are proposed, and such services are characterized by a myriad of constraints and advantages. These services can be broadly be categorized into centralized (range-free) and distributed (range-based) [9]. The network may consist of homogeneous and heterogeneous nodes to achieve the goal. Accordingly, the localization strategies vary, to account for different quality of service (QoS) parameters and application requirements. In other terms, the localization process, particularly a 2D localization process, can be likened to a distance estimation process among the nodes. Commonly employed techniques for distance estimation include received signal strength indicator (RSSI), the angle of arrival (AOA), time of arrival (TOA), time difference of arrival (TDOA), angulation, and lateration, among others [9].

As mentioned earlier, AHP enables decision-makers to structure complicated problems as a hierarchical structure. AHP is successfully employed to do multi-criteria decision-making for a wide range of engineering applications [10–12], ranging from a car purchase, assessing the viability of industrial projects, selecting store locations, selection of certain Six Sigma projects, market analyses, information systems re-engineering assessment, power system, and software reliability allocation, etc. [13–15]. The service selection process utilizes a sufficient number of criteria to select the service, and the multi-criteria decision analysis method (MCDA) needs analysis in detail [16]. Also, AHP may be applied where the decision is based on criteria, but it is not inter-dependent during the decision dependency.

Figure 10.1 depicts an AHP hierarchy of alternatives, criteria, and goal. The alternatives find the suitable value for the criteria, using minor issues, and so numerous problems may be formulated as AHP hierarchy, using graph transforms. All elements may not directly interact with and affect each other, but sometimes elements may interact with each other indirectly. Figure 10.2 shows the generic steps for multi-criteria decision-making using AHP.

This chapter presents an AHP-based service selection and composition scheme for localization services in WSN. The following section presents the service selection schemes for WSN localization. The discussions are based on the FSONA framework.

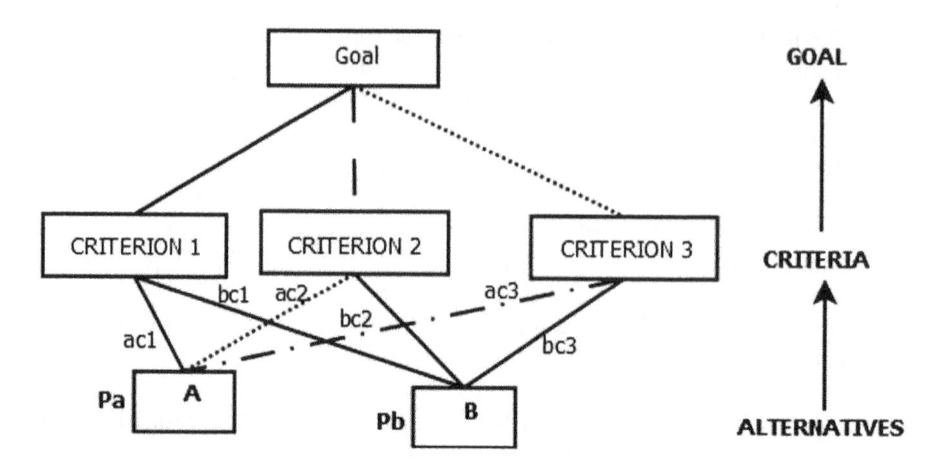

FIGURE 10.1 AHP hierarchical structure.

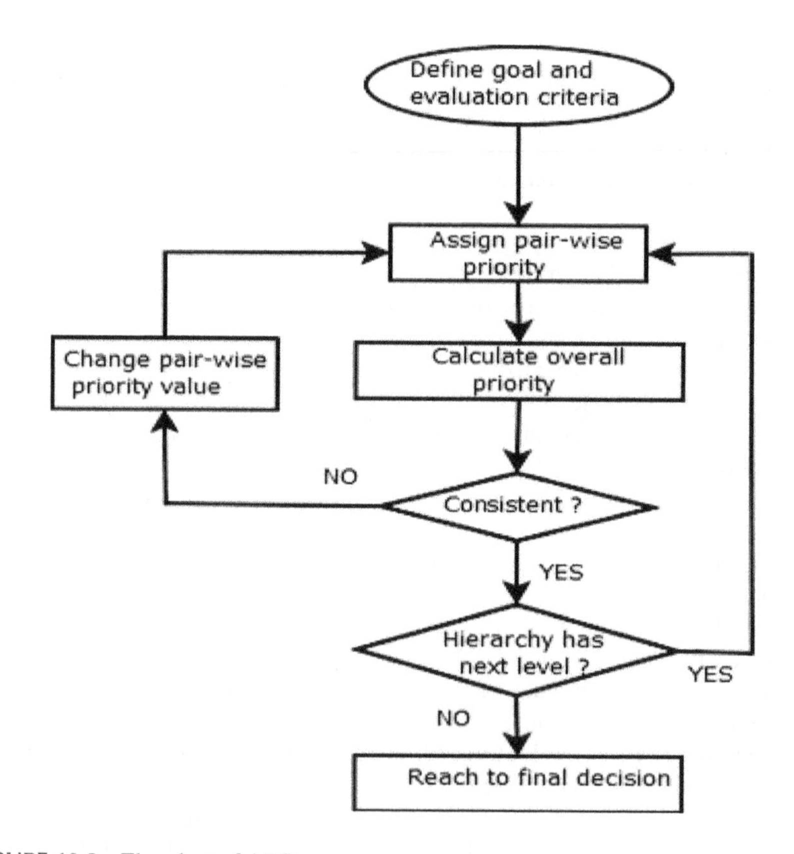

FIGURE 10.2 Flowchart of AHP.

10.3 FSONA-BASED LOCALIZATION SERVICES

To fulfill application requirements, services that are available for a particular functionality, such as WSN localization, are growing at a rapid rate. For instance, commonplace services include APIT, DV–hop, Improved DV-hop, Centroid, MDS MAP, IMDS-MAP, ADAL, TOA, TDOA, and so on [17–21]. The available services have QoS parameters, such as accuracy, communication cost, computation cost, node deployment, hardware cost, etc., which are required during network setup. It is difficult for a user to select an appropriate service from the available services pool without the knowledge of the low-level details. The identification of low-level information is a time-consuming and complex task.

At present, service selection and composition is a challenging task, as there are multiple numbers of services available on the web or cloud. These services are developed based on different QoS parameters and standards. Although the services are available for the task, the selection of service is difficult for consumers, due to unawareness of QoS parameters. Identification of QoS parameters is also one of the tedious tasks for the comparison of services. Different consumers have different requirements; for example: for a consumer, it is possible that the accuracy of the event location is more important than the image of the event. On the other hand, for a different consumer, the image of the event is more important than the accuracy of the event location. In a case where two QoS parameters are relatively crucial is traffic monitoring. The image and motion sensor nodes are deployed on the roadside to capture the image and movement of vehicles. The goal of this application is to detect the accurate place of an accident if it occurs, which is more important than the image of vehicles and communication cost. The image and communication cost is relatively less important than accuracy. Therefore, importance and QoS parameters may differ as per the requirements and constraints of applications. In the best service selection, comparisons of QoS parameters are required. These QoS parameters may be categorized into different categories and subcategories at different levels.

10.3.1 Components of the FSONA Framework

In the FSONA framework, service composition is a process to make the combination of different functionalities as per the application's requirements. The process is to generate the service graph, using different combinations of service components. A service graph explains the execution flow of functionalities.

BB Description: A building block (BB) is a basic module of the composition process. In the composition process, the complete description, i.e., constraints and minimum requirements, must be followed.

Requirement Description: To smoothly perform the composition process, requirement specification should be harmonized with the offered BB description.

Identification of Dependencies: Dependency is a major challenge in the functional composition process. To simplify the composite service, it is necessary to understand how different BB or service components depends and affects others.

Service Rating: It is complicated to identify the service suitability when various services provide the same functionality. In FSONA, qualitative and quantitative attributes are defined in the service, but the identification of appropriate comparison methods of the same service is a critical task.

The service selection and composition model is developed based on AHP. The proposed model is used to select the most suitable service among the available services to localize the sensor nodes, which is based on the QoS parameters and its ranking.

10.3.2 Formalization of FSONA Services

The proposed architectural framework concedes services as a collection of visual effects, which are defined as the property of service. There are two kinds of properties to define the service, i.e., Mandatory and Additional properties. The Mandatory properties (M) of service are described as follows [7]:

- These are the guaranteed properties.
- It does not hold any type of service rating.
- This property is used to determine whether the service has a precise property or not, for example, localization, etc.
- The Mandatory properties represent the necessary properties in a requirement specification.

Additional properties (A) of service are explained below [7]:

- The additional property is presented using a qualitative property.
- These properties are not guaranteed for the user's application.
- It holds the service rating explicitly.
- Qualitative properties are useful for the required properties by the users, and they are described in the requirement specification, for example, accuracy, computation power, error rate, random deployment, or distributed deployment, etc.

In the process of service selection and composition, the user inserts the request as an input with the Mandatory and additional properties, using a predefined interface to obtain the service, as per the application requirements. After that, formalization is needed to perform the service selection and composition with predefined specifications. Once the standard is achieved, the service registry executes the matching process with the available service list and returns to the user, based on the requirement and availability of service.

The user requests (*Ur*) for the service, and services may be offered by the service provider (*Po*). So the user requests (*Ur*) consist of two requirements (*Mr, Ar*) and the service provider (*Po*) holds the same property (*Mo, Ao*). Additional property must hold the service rating value (*q*) in the offered specification for enabling the comparison in the same kind of available service. The value is employed $q \in [0, 1]$ in comparison. The requested specification contains the A along with specified properties (*p*) of the user. The format of the user request is formalized as:

$$Ur = (Mr,\ Ar, p) \tag{10.1}$$

The offered specification with additional properties is formalized as:

$$Po = (Mo,\ Ao, q) \tag{10.2}$$

And the offered specification contains the additional properties along with their rating values, which are expressed as:

$$So = (Mo,\ ko, q) \tag{10.3}$$

$$Ao = (Aod,\ Anod) \tag{10.4}$$

Where *Aod* is the properties which are defined in the requirement specification, and *Anod* is not defined in the specification. As per the nature of the requested service:

- It is determined whether mandatory properties have been satisfied with the offered service or not and whether offered services are suitable or not for the requester.
- Along with mandatory properties, network constraints are also considered with additional properties and priorities to offer the best service as per the given requirement and offer.

In the implementation, constraints are considered, which are: DE = Deployment, ND = Node Density, OB = Obstacle, NM = Node Mobility, MA = Mobile Assisted, CC1 = Communication Cost, CC2 = Computation Cost, HW = Hardware Cost, ACC = Accuracy, and Anchor Node = AN [10, 11].

The goal is to obtain the best service as per the requirement and offer. The maximum priority of the additional property is considered. The objective function is represented as:

$$Max \sum_{i=0}^{n} (pi * qi) \tag{10.5}$$

The notion is to attain the best service and maximize the values having higher priority and additional properties of a service. Therefore, the objective function is expressed as:

Subject to constraints:

1. DEod ≤ DEr
2. NDod ≤ NDr
3. OBod ≤ OBr
4. CC1od ≤ CC1r
5. CC2od ≤ CC2r
6. NMod ≤ NMr
7. MAod ≤ MAr
8. ANod ≤ ANr
9. HWod ≤ HWr
10. ACCod ≤ ACCr

Where (DEr, NDr, OBr, CC1r, CC2r, NMr, MAr, ANr, HWr, ACCr) are requesters parameters and (DEod, NDod, OBod, CC1od, CC2od, NMod, MAod, ANod, HWod, ACCod) are offered by the service provider.

The rating values are considered based on the rating scale, which is already used [13] to rate the qualitative properties of service. Based on the rating values given by the users, services are offered. The scale of rating defines the highest and lowest rate as per the value that is defined from (0 … 9).

10.3.3 Necessities of Service Selection

The service selection process is implemented with consideration of application requirements and constraints of users. In this process, the application sends their constraints and requirements to the service broker. Subsequently, the broker performs the service-matching process with the priority of different criteria that are given by the user. Finally, the application obtains the most suitable offered service. Some necessities are needed to perform the matching process. The necessities are explained below.

Service is the collection of qualitative and inherited properties that is described as:

> S = <B, C>, where B, C are inherited and qualitative property, respectively. The visual effect of specific service is described by the service description, and these properties are required to perform the comparison based on the requirements.

The assumption is that two services that are Si and Sj provide the same functionality with different constraints and requirements. Hence, the services can be compared, based on the knowledge of requirement and condition as follows:

- The definition of service should be precise and clear.
- How the best and appropriate service can be differentiated during the finding.

In order to consider the service, this may be either requested by a user (Ur) or service provider (P0), and presented as: Sr = <Br, Cr> and S0 = <B0, C0>. Comparison between Si and Sj is given in Table 10.1(a). To get the optimal result as per the constraints, Ur and P0 are compared, and the results are listed in Table 10.1(b).

At present, various service providers are providing similar services to facilitate the users and applications. The users are growing with the new demands for different applications. In the present scenario, the services may be limited, but in the coming future, the number of services is going to increase significantly and is going to be provided by the different service providers with different constraints and implementation scenario. The identification of service is a time-consuming and challenging task without the knowledge of technical details. Therefore, the service-matching process plays a major role.

TABLE 10.1(a)
Service Comparison

Service Si and Sj	Conditions
Equivalent	Bi = Bj
Identical Sj is a refinement of Si	<Bi, Ci> = <Bj, Cj>
Sj is a refinement of Si	Bi $\subset$ Bj

TABLE 10.1(b)
Service Matching

Condition	Result
Sr < S0	User needs may not be fulfilled
Sr > S0	User needs may not be fulfilled
Sr = S0	Precise match
S0 ≠ Sr	No match and user can't proceed

10.3.4 Service Composition Using BB and Workflow

The service composition is a way to build a composite service. It describes the interaction pattern between service modules to provide the service with specific features required by the user using the workflow. Some necessities are needed for the composition process when the user sends the requirements for novel service that may not be fulfilled efficiently with the existing service, or that does not exist in the list of available services. So, to build the workflow, the necessary steps are required, such as:

- Suitable BBs or services should be selected to fulfill the requirements that suits.
- These selected BBs or services should be arranged and linked with a specific pattern to give the absolute service as a response to the request.

The composition process adapts the requirements and constraints of the application. This is required when a single service cannot fulfill the need of service request completely, but a group of BBs or services may. It is performed by achieving the global or local optimal solution. To consider the requirements, i.e. application requirements and network constraints, the dynamic composition is performed to satisfy the request (as depicted in Figure 10.3).

10.4 AHP-BASED SYSTEM MODEL

To perform the service selection and composition with a system, the following properties are considered:

- In the setup, all nodes are homogeneous and suitable software is used to deploy.
- Nodes may act as a service requester.
- The service provider strives to meet the request.
- The requester requests for service with propagation channel, using a hop-by-hop pattern.

The communication links are established in the ideal mode.

Let "*App*" be the set of applications which are supported by the FSONA [7]. The application set is presented as:

$$App \in \{app1, app2, app3, app4, ... appn\} \tag{10.6}$$

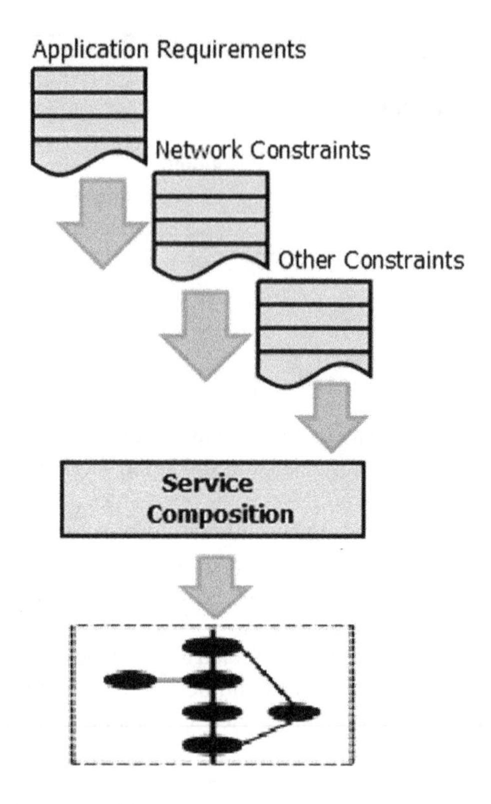

FIGURE 10.3 Service composition.

Applications carry different information, which is generally categorized into two categories, i.e., mandatory (M) and additional information (A). Where mandatory information (M) and additional information (A) are expressed as follows:

$$M \in \{m1, m2, m3, m4, ,\ldots\ldots, mn\} \tag{10.7}$$

$$A \in \{a1, a2, a3, a4, \ldots\ldots, am\} \tag{10.8}$$

In the case of "k = 0", the interaction of an application is similar to the current application, and the user gets the minimum facility in service selection because of the only mandatory requirement. If "k = 1", the application is coarsely tuned, and as the number of parameters increases with k, the application is better tuned and filtered in the selection process. The application utilizes different services for localization as per the parameters, and these parameters play a significant role in the selection of the most suitable service for a different deployment scenario.

Let "S" be the total number of available service and represented as:

$$S = \{s1, s2, s3, s4, \ldots.\ sm\} \tag{10.9}$$

An ordered service can be described as the set of services which are used by an application App. The ordered set is presented as: $S_k = \{s_{k1}, s_{k2}, s_{k3} \ldots. s_q\}$, where $s_k i > s_k i + 1$, $app_k = s_k \mid s_k \subset s\}$ i.e., Centroid, DV hop, APIT, etc., which are used for its functionality. In addition to this, S_f is a set of services that supports a specific application app_f to provide the service efficiently.

So, the set of supported services are expressed as:

$$Sf = \{sf1, \text{sf}2, sf3 \ldots. sfm\} \tag{10.10}$$

where $a_f = \text{s}_\text{f} \mid \text{s}_\text{f} \subset \text{S}$ and $\text{S} \supset \text{s}_\text{f}$ and universal set of services "*Su*" which consists of all kinds of services are represented as:

$$Su = \{S1U \ldots.UScentroid, USDV - hop, USAPIT\ U \ldots\ldots Un\} \tag{10.11}$$

where US describes the universal set and all services are of an ordered subset.

If S_i follows the service format "abc," then it is written as $Si = Sabc$, and similarly, other services also may be presented with this example: S_k = SAPIT ∪ SPIT, which means the complete suite of APIT.

User preferences as a request are formularized as a process of converging the service requests. The request contains the criterion and subcriterion for getting the best available service. Therefore, pointwise convergence is performed for each criterion and subcriterion. The pointwise convergence gives the service as a result, as per each criterion or subcriterion. The almost sure convergence shows that process of resultant service, which may not fulfill all preferences as quantified.

10.4.1 Convergence of Service Request

$$\text{If}\quad \text{E}\,(\text{X} - App)2 = 0 \tag{10.12}$$

Then X = 'A' with probability 1 or X converges to S with probability 1. Requester wants to get the application with a tolerable limit of ϵ choice, which does not satisfy a few requests of:

$$\sum_{k=1}^{m} Ak \tag{10.13}$$

Let $\sigma^2 x$ be the measurement variability; according to Chebyshev inequality, it is shown as:

$$\text{S}\{\text{X} - App \mid \leq \epsilon\} \geq 1 - \sigma 2x / \epsilon 2 \tag{10.14}$$

If $\sigma x <<< \epsilon$, then σx is not small enough than ϵ, then the results are not accurate. To improve the accuracy of the result, "*App*" needs to take n measurements related to n random variables $\{X_i = 1.........., n\}$, mean "*App*" and noise random variable W_i, calculated by:

$$Xi = App + Wi, \; i = 1.....,n \tag{10.15}$$

The average of the random variable "*n*" is shown as:

$$\mathrm{X} = (X1 + X2 + X3 + ... + Xn)/\mathrm{n} \tag{10.16}$$

where X, $\sigma^2 x$ is a mean of *S* and variance, respectively. According to inequality of Chebyshev, this is shown as:

$$\mathrm{S}\{|\; \mathrm{X} - App \;| < \epsilon\} \geq 1 - \sigma 2x/\mathrm{n}\epsilon 2 \tag{10.17}$$

10.4.2 Point-Wise Convergence of Service Request

The discrete order of random variables is described as $R_vX_1, R_vX_2, R_vX_3, \ldots R_vX_n$ and converges to limiting random variable R_vX, if and only if for any $\epsilon > 0$; however, for smaller, find n_0 so that $|\; R_vX_n\,(\xi)$- $\mathrm{R_vX}(\xi)| < \epsilon \;\forall\; n > n_0$ and ξ, it is right for variables but highly restrictive for random variables.

10.4.3 Almost Sure Convergence of Service Request

A sequence of random variables $\{R_vx_n\}$ converges definitely to random variable R_vx, when every ξ point in the similar space S fulfils the following criterion: $Lim\; n \rightarrow \infty \;|\; R_vX_n\,(\xi) - R_vX(\xi)\;| < \epsilon \rightarrow 0\; n \rightarrow \infty$ with probability 1, and it may also be symbolized as:

$$S(Rvxn \rightarrow Rvx) = 1 \text{ as } n \rightarrow \infty \tag{10.18}$$

10.5 IMPLEMENTATION, EXPERIMENTATION, AND RESULTS

The implementation process consists of four modules:

a. Criteria identification, which describes the performance evaluation criterion of localization as a service.
b. Creation of a hierarchical model presents the hierarchical structuring of different criterias and subcriteria of services as a QoS parameters.
c. Pairwise comparisons are performed among the QoS parameters based on their rating preferences.
d. Identification of final priority and decision is obtained as per the QoS parameters and its local and global priorities of services.

10.5.1 Criteria Identification

In the implementation process, the first task is criteria identification. The criteria identification for each localization service in the WSN is a challenging task, as for all the services, performance criteria is not defined.

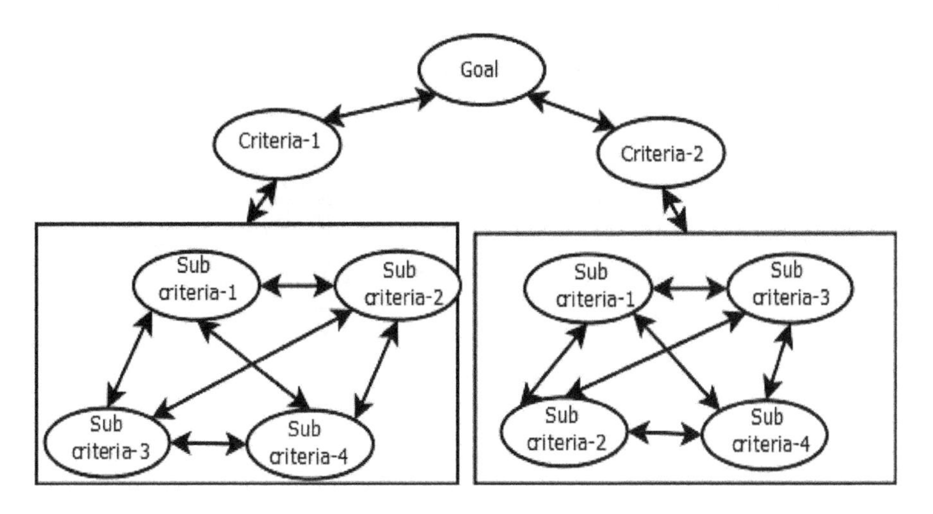

FIGURE 10.4 Structural division of criteria.

But the detailed study criterion are identified as: i.e., DE = Deployment, ND = Node Density, OB = Obstacle, NM = Node Mobility, MA = Mobile Assisted, CC1 = Communication Cost, CC2 = Computation Cost, HW = Hardware Cost, ACC = Accuracy, AN = Anchor Node, and implemented as an application and network requirements. These are implemented using the GUI and API, where criteria are implemented for the comparison. Every criterion is compared with its category. The method of comparison is depicted in Figure 10.4 to achieve the goal.

10.5.2 Creation of Hierarchical Model

Building the hierarchical structure of criteria is the second step for service selection and composition. It requires an in-depth knowledge of various criteria and subcriteria to select the best service for localizing the sensor nodes. The hierarchy is designed such that all alternatives can satisfy the criteria, and subsequently, the main goal. We divide the given problem into the AHP model as discussed in a systematic manner and thus create a hierarchical model, shown in Figure 10.5.

10.5.3 Pair-Wise Comparison

A pair-wise comparison is performed to prioritize the criteria and subcriteria. Pair-wise comparison among the various criteria and subcriteria is depicted in Figure 10.5. It also shows the pair-wise comparison among all the available service (alternatives) for each subcriterion. To build the hierarchical model, the pair-wise comparison matrix is developed from the given criteria and subcriteria with the help of the preference table defined by the AHP process, as shown in Table 10.2. Initially, the pair-wise comparison matrix is built for all available services with respect to each subcriterion.

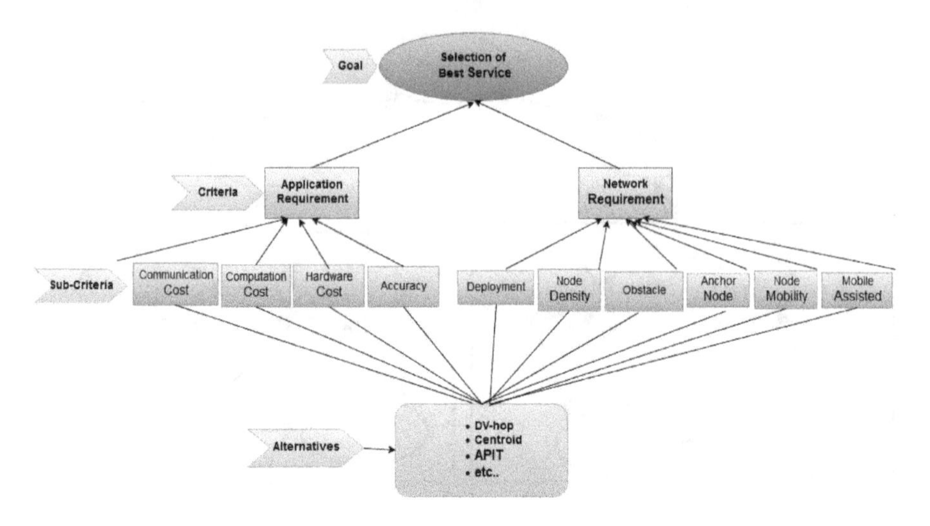

FIGURE 10.5 AHP for the service selection case.

Also, the matrices remain fixed as each alternatives and performance is not affected concerning certain subcriteria. As these matrices remain fixed, they are known as Developer Matrix. The following steps are required to determine priorities from the pair-wise comparison matrix:

- Calculate the sum of each column and let's say s[i]. item. Normalize each column by dividing each value in a column with the sum of the respective columns.
- Determine the sum of each row; this sum gives the weight column vector, with the help of prioritized values.

It is assumed that the developer has rated the service based on the performance as per the specified criteria. In the architectural framework FSONA [7], the following services are implemented and treated as alternatives, which are used to localize the sensor nodes.

- Distance Vector hop (DV-hop)
- Centroid
- APIT

10.5.4 Identification of Resultant Priority and Decision

In this step, the priority of criteria and each subcriterion based on the user preferences are identified. Two cases are considered for testing the correctness of the service selection process. These cases are based on the different requirement preferences of users for different applications. See Figure 10.6 for a comparison. See Tables 10.2 to 10.4 for further information.

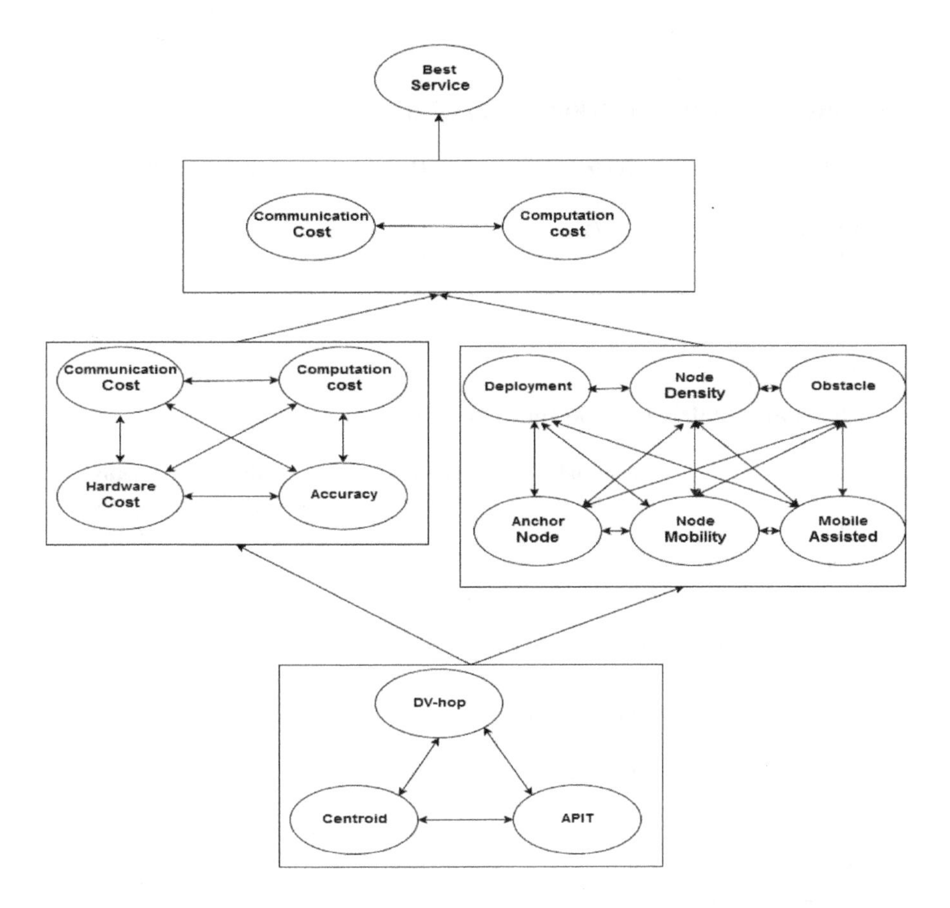

FIGURE 10.6 Pairwise comparison.

10.5.5 Results and Discussion

In the first case, it is assumed that the application requirement is more important than the network requirement. Therefore, the user using the GUI, which is shown in Table 10.13, has given the priority. See Tables 10.5 through Table 10.10 for further information.

TABLE 10.2
Developer Matrix (with Respect to CC1)

CC1	Centroid	DV Hop	APIT	Priority
Centroid	1	1/7	1/4	.30363
DV hop	7	1	7	.80126
APIT	4	1/7	1	.33669

TABLE 10.3
Developer Matrix (with Respect to CC2)

CC1	Centroid	DV Hop	APIT	Priority
Centroid	1	4	4	.66666
DV hop	1/4	1	7	.16666
APIT	1/4	1	1	.16666

TABLE 10.4
Developer Matrix (with Respect to CC2)

CC1	Centroid	DV Hop	APIT	Priority
Centroid	1	1/3	1/5	.12386
DV hop	3	1	1/2	.35200
APIT	5	2	1	.61357

TABLE 10.5
Developer Matrix (with Respect to Accuracy)

CC1	Centroid	DV-Hop	APIT	Priority
Centroid	1	1/4	1/7	.33421
DV-hop	4	1	1	.33421
APIT	7	1	1	.49821

TABLE 10.6
Developer Matrix (with Respect to Deployment)

CC1	Centroid	DV-Hop	APIT	Priority
Centroid	1	3	1	.42857
DV-hop	1/3	1	1/3	.14825
APIT	1	3	1	.42857

TABLE 10.7
Developer Matrix (with Respect to Node Density)

CC1	Centroid	DV-Hop	APIT	Priority
Centroid	1	1/4	1/4	.10846
DV-hop	4	1	4	.62433
APIT	4	1/4	1	.26719

TABLE 10.8
Developer Matrix (with Respect to Obstacle)

CC1	Centroid	DV-Hop	APIT	Priority
Centroid	1	1	1	.33333
DV-hop	1	1	1	.33333
APIT	1	1	1	.33333

TABLE 10.9
Developer Matrix (with Respect to Anchor Node)

CC1	Centroid	DV-Hop	APIT	Priority
Centroid	1	1	3	.42857
DV-hop	1	1	3	.42857
APIT	1/3	1/3	1	.14285

An application requirement consists of various subcriteria, such as Communication Cost, Computation Cost, Hardware Cost, and Accuracy. Further, the preference of each subcriteria is given by the user, using a pairwise input method. The input and priority are determined and shown in Table 10.12. The network requirement also consists of subcriterion Deployment, Node Density, Obstacle, Anchor Node, Node Mobility, and Mobile-Assisted-based. Similar to the application requirement, the preferences are assigned to subcriteria, and a priority matrix of network and network requirement is achieved, which is depicted in Table 10.17. Then a resultant priority is obtained, and a criteria weight tree is formed, which shows the criteria, subcriteria, and alternatives, along with their respective priorities, as shown in Figure 10.6. The priorities are described as a Local Priority (L) and Global Priority (G), which are defined below:

- **Local Priority:** It is the priority of a node in comparison to its parent node.
- **Global Priority:** It is the priority of a node with respect to its root (goal) node.

TABLE 10.10
Developer Matrix (with Respect to Mobile Assisted)

CC1	Centroid	DV-Hop	APIT	Priority
Centroid	1	1	1/4	.17276
DV-hop	1	1	1/3	.19374
APIT	4	3	1	.63681

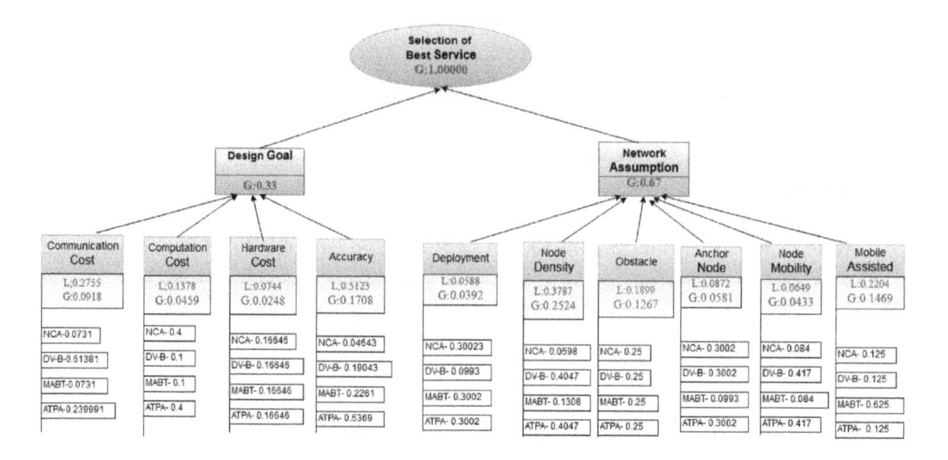

FIGURE 10.7 Criteria weight tree for case 1.

These priorities are used in the criteria weight tree to describe the local and optimal solution. All the priorities determined from the pair-wise comparison matrices are local. After that, global priority for each node is identified using the following defined formula:

$$\text{Global Priority (of a node)} = \text{Local priority of the node} \\ * \text{(Global priority of parent node)} \tag{10.19}$$

Figure 10.7 shows the criteria weight tree for case 1.

For example, the local priority of accuracy is 0.5123 for application requirement, and the global priority of accuracy for the root node (goal) is 0.1708. All the priorities calculated from the pair-wise comparison matrices are local priorities. We thus calculate the global priority for each node with the formula:

$$\text{Resultant priority of the alternatives} = \sum_{1}^{n} \textit{Priority of alternatives} \\ \textit{w.r.t.Criterian} * (\textit{Global Priority of Criteria} \ldots n) \tag{10.20}$$

Figure 10.8 shows the criteria weight tree for case 2.

Resultant Priority of Centroid = (Priority of Centroid with respect to DE*Global Priority of DE) + (Priority of Centroid with respect to ND * Global Priority of ND) + (Priority of Centroid with respect to OB * Global Priority of OB) + (Priority of Centroid with respect to NM * Global Priority of NM) + (Priority of Centroid with respect to MA * Global Priority of MA) + (Priority of Centroid with respect to CC1 * Global Priority of CC1) + (Priority of Centroid with respect to CC2 * Global Priority

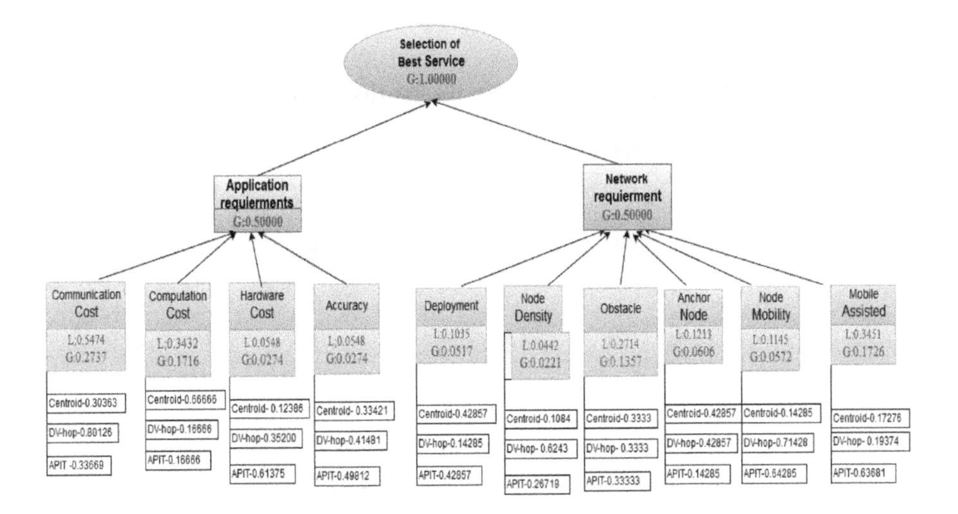

FIGURE 10.8 Criteria weight tree for case 2.

of CC2) + (Priority of Centroid with respect to HW * Global Priority of HW) + (Priority of Centroid with respect to ACC * Global Priority of ACC). QoS parameter weights are quantified as per the user preference for case 1 and case 2. Global preferences are quantified in Tables 10.11 and 10.15 for case 1 and case 2, respectively. Local preferences are quantified from Tables 10.12 and 10.13 and Tables 10.16 and 10.17 for case 1 and case 2, respectively. These tables are required to find the priority

TABLE 10.11
User Preference Matrix for Case 1

Requirements	Application Requirements	Network Requirements	Priority
Application	1	2	.67
Network	1/2	1	.33

TABLE 10.12
User Preference Matrix for Subcriteria: Application Requirement

Criteria	CC1	CC2	Hardware Cost	Accuracy	Priority
CC1	1	2	4	1/2	.27553
CC2	1/2	1	2	1/4	.13776
Hardware cost	1/4	1/2	1	1/6	.07436
Accuracy	1/2	4	6	1	.51233

TABLE 10.13
User Matrix Preference for Subcriteria: Network Requirement

Criteria	De	Nd	Ob	An	Nm	Ma	Priority
De	1	1/5	1/3	1	1/3	1/3	.05884
Nd	5	1	3	4	7	2	.37867
Ob	3	1/3	1	2	6	1	.18998
An	1	1/4	1/2	1	3	1/3	.08718
Nm	3	1/7	1/6	1/3	1	1/6	.06490
Ma	3	1/2	1	3	6	1	.22039

TABLE 10.14
Resultant Priorities of the Alternatives (Case 1)

Alternatives	Resultant Priority
DV-hop	.4504
Centroid	.26126
APIT	.39493

TABLE 10.15
User Preference Matrix for Case 2

Requirements	Application Requirements	Network Requirements	Priority
Application	1	1	.5
Network	1	1	.5

TABLE 10.16
User Matrix Preference for Subcriteria of Application Requirement Case

Criteria	CC1	CC2	Hardware Cost	Accuracy	Priority
CC1	1	2	2	9	.54733
CC2	2	1	7	7	.34311
Hardware cost	1/9	1/7	1	1	.05477
Accuracy	1/9	1/7	1	1	.05477

TABLE 10.17
User Matrix Preference for Subcriteria of Network Requirement

Criteria	De	Nd	Ob	An	Nm	Ma	Priority
De	1	5	1/3	1	1/3	1/3	.10347
Nd	1/5	1	1/3	1/3	1/5	1/5	.04421
Ob	3	3	1	2	6	1	.27143
An	1	3	1/2	1	3	1/5	.12127
Nm	3	5	1/6	1/3	1	1/6	.11446
Ma	3	5	1	5	6	1	.34513

TABLE 10.18
Resultant Priorities of the Alternatives (Case 2)

Alternatives	Resultant Priority
DV-hop	.43553
Centroid	.43565
APIT	.37449

of service with respect to criterion (QoS parameters). Table 10.14 and Table 10.18 represent the resultant ranking value for cases 1 and 2, respectively.

The final ranking of service for local and global criterion weight for case 1 is depicted in Figure 10.9, and similarly, for case 2 in Figure 10.10. The ranking of service is obtained based on Tables 10.15 to 10.17. In case 1, as mentioned in

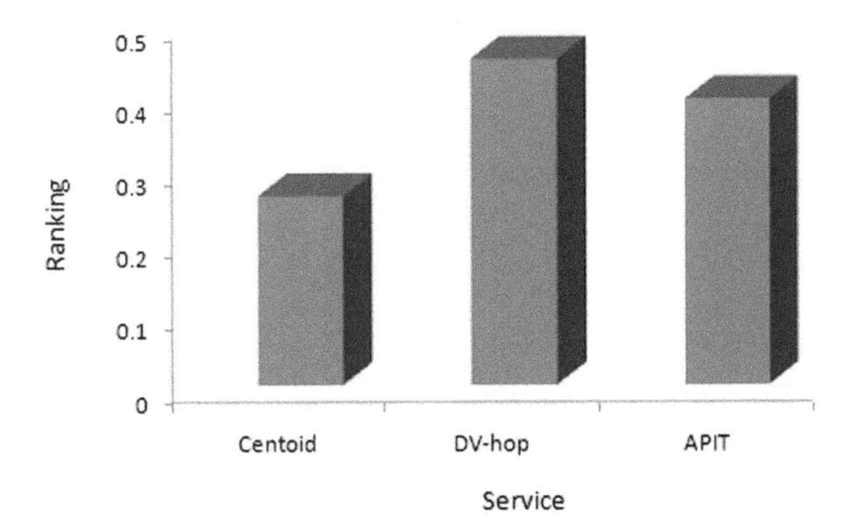

FIGURE 10.9 Final ranking of the services for case 1.

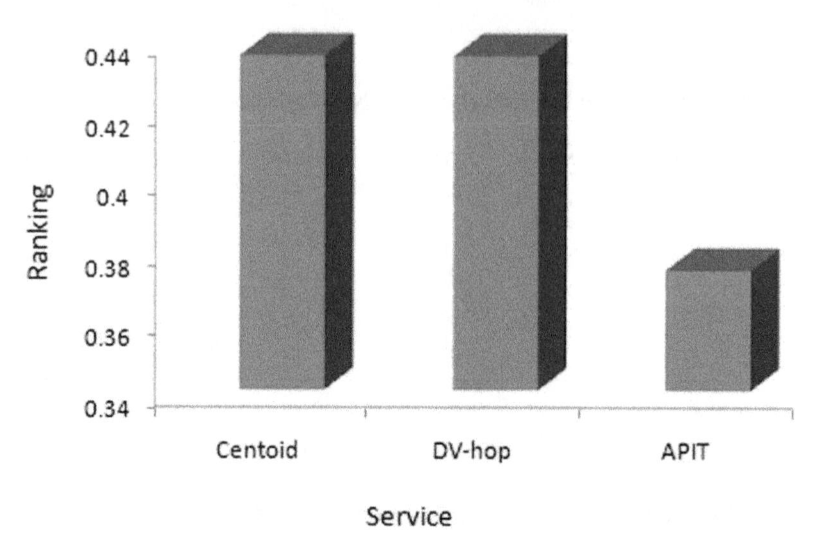

FIGURE 10.10 Final priorities of the services for case 2.

Table 10.14, DV-hop is the best-ranked service as per the given criterion. And for case 2, centroid is ranked as the best service as per the given criterion. Figure 10.11 shows the number of comparisons as per the number of criteria. With the increase of criterion, the number of comparisons shall increase based on the factor $(n * n - 1)/2$. The local priority of services for criterion and weight is depicted in Figure 10.12.

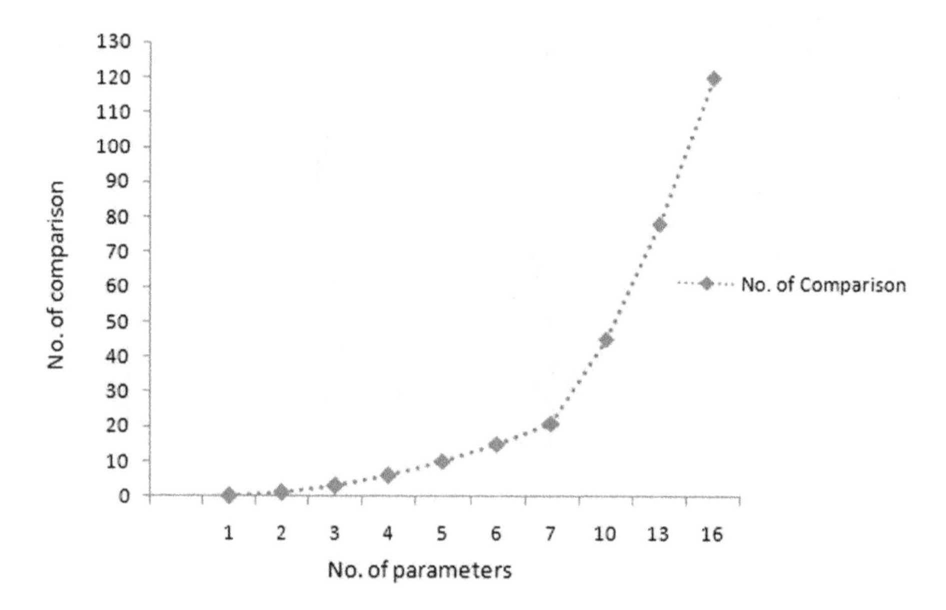

FIGURE 10.11 Required number of comparisons based on the criteria.

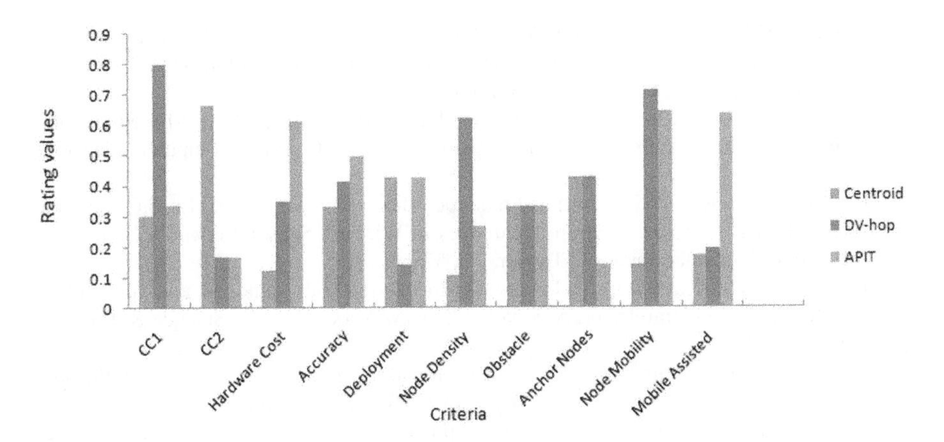

FIGURE 10.12 Service with respect to criteria.

For the pair-wise comparison, the criteria weight tree is depicted in Figures 10.7 and 10.8 for case 1 and case 2, respectively.

10.6 CONCLUSIONS

This chapter has presented an objective way of assigning priorities to different services for the same software functionality, based on an AHP-based multi-criteria decision for localization service in WSN. Subjective preferences, such as network and application requirements, have been quantified by the proposed AHP model. Case studies have exemplified, and the analysis has accounted for, a number of QoS parameters, such as accuracy, cost, node mobility, etc. Based on requirements provided by users, different services are ranked differently, demonstrating that AHP can successfully amalgamate functional requirements with users' subjective requirements. In the future, we plan to extend this research work with more numbers of services and criterion (QoS parameters) for varied applications.

REFERENCES

1. Perrey, Randall, and Mark Lycett. "Service-oriented architecture," Proceedings in 2003 Symposium on Applications and the Internet Workshops, IEEE, 2003.
2. Laskey, Kathryn B., and Kenneth Laskey. "Service oriented architecture," Wiley Interdisciplinary Reviews: Computational Statistics 1, no. 1(2009): 101–105.
3. Zhu Ming, *et al.* "An approach for QoS-aware service composition with GraphPlan and fuzzy logic." Procedia Computer Science 141(2018): 56–63.
4. Erl Thomas. Service-oriented architecture: concepts, technology, and design. India: Pearson Education, 2005.
5. Duggan, Dominic. "Service-Oriented Architecture." Enterprise Software Architecture and Design: Entities, Services, and Resources (2012): 207–358.
6. Saaty, Thomas L. "Decision making with the analytic hierarchy process." International Journal of Services Sciences 1, no. 1(2008): 83–98.

7. Singh, Akhilendra Pratap, O. P. Vyas, and Shirshu Varma. "Flexible service oriented network architecture for wireless sensor networks." International Journal of Computers Communications & Control 9, no. 5 (2014): 610–622.
8. Zahra Taghikhaki, Nirvana Meratnia, and Paul J. M. Havinga. "A trust-based probabilistic coverage algorithm for wireless sensor networks." Procedia Computer Science 21 (2013): 455–464.
9. Santar Pal Singh. (2015). Author links open the author workspace. In: S.C. Sharma, (ed.), Range Free Localization Techniques in Wireless Sensor Networks: A Review. Procedia Computer Science, Elsevier, vol. 57, pp. 7–16.
10. Saaty, Thomas L., and Luis G. Vargas. Models, methods, concepts and applications of the analytic hierarchy process, vol. 175. New York: Springer Science & Business Media, 2012.
11. Saaty, Thomas L. Fundamentals of decision making and priority theory with the analytic hierarchy process, vol. 6. Pittsburgh, PA: RWS Publications, 2000.
12. Saaty, Thomas L. The analytic hierarchy process: planning, priority setting, resources allocation. New York: McGraw, 1980.
13. Saaty, Thomas L. "Decision making—the analytic hierarchy and network processes (AHP/ANP)." Journal of Systems Science and Systems Engineering 13, no. 1 (2012): 1–35.
14. Roy, Diptendu Sinha, Dusmanta Kumar Mohanta, and A. K. Panda. "Software reliability allocation of digital relay for transmission line protection using a combined system hierarchy and fault tree approach." IET Software 2, no. 5 (2008): 437–445.
15. Selwa Elfirdoussi and Zahi Jarir. "An integrated approach towards service composition life cycle: A transportation process case study." Journal of Industrial Information Integration 15(2019): 138–146.
16. Hamed Bouzary, F. Frank Chen, and Krishnan Krishnaiyer. "Service matching and selection in cloud manufacturing: a state-of-the-art review," Procedia Manufacturing 26(2018): 1128–1136.
17. J. Kuriakose, V. Amruth and N. S. Nandhini. (2014). A SURVEY on localization of Wireless Sensor nodes, International Conference on Information Communication and Embedded Systems (ICICES2014), Chennai, India, pp. 1–6.
18. Tashnim J.S. Chowdhurya, Colin Elkina, Vijay Devabhaktunia, Danda B. Rawat, Jared Oluochc. (2016). Advances on localization techniques for wireless sensor networks: A survey, Computer Networks, Elsevier, vol. 110, pp. 284–305.
19. Amundson I., Koutsoukos X. D. (2009). A survey on localization for mobile wireless sensor networks. In: Fuller R., Koutsoukos X. D. (Eds.). Mobile Entity Localization and Tracking in GPS-less Environments. Lecture Notes in Computer Science, vol. 5801. Springer, Berlin, Heidelberg. Terrain Part I—Theory and modelling. IEEE Transactions on Systems, Man, and Cybernetics, Part C: Applications and Reviews, 41(3), pp. 376–382.
20. Subir Halder, Amrita Ghosal. "A survey on mobility-assisted localization techniques in wireless sensor networks." Journal of Network and Computer Applications 60(2016): 82–94.
21. Leila Chelouaha, Fouzi Semchedine, Louiza Bouallouche-Edjkounea. "Localization protocols for mobile wireless sensor networks: A survey." Computers & Electrical Engineering, Elsevier 71(2018): 733–751.

11 Catalyst Is Important Everywhere

The Roles of Fog Computing in an IoT-Based e-Healthcare System

Sankalp Nayak and Debajyoty Banik
Kalinga Institute of Industrial Technology, Bhubaneswar, India

CONTENTS

11.1 Introduction ... 195
11.2 Architecture and the Role of Fog Computing in IoT Healthcare System ... 199
11.2.1 IoT e-Health Ecosystem Architecture ... 199
11.2.1.1 IoT e-Health Device Layer ... 200
11.2.1.2 IoT e-Health Fog Layer ... 202
11.2.1.3 IoT e-Health Cloud Layer ... 206
11.2.2 IoT e-Health Challenges ... 209
11.3 Case Study ... 211
11.3.1 Health Monitoring System Using Early Warning Score: Case Study Based on Internet of Things Using Fog Computing ... 211
11.3.1.1 Sensor Network ... 214
11.3.1.2 e-Health Gateway ... 214
11.3.1.3 Backend Layer ... 218
11.4 Conclusions ... 218
References ... 219

11.1 INTRODUCTION

Nowadays, the use of hardware implementation [1–7], sensor network [8], and Internet of Things (IoT) are remarkably increased. The IoT is considered as a dynamic global network infrastructure and an ever-growing ecosystem that unifies hardware, computing devices, physical objects, and people to interact, communicate, collect, and exchange data over a network. With the introduction of IoT to healthcare, the quality and the cost of medical care have been improved, primarily due to the computing nature of IoT. It automates the continuous monitoring of all the healthcare systems, which were previously performed by humans. With the rapid development of IoT healthcare systems, there is a significant growth in deployment of the large number of sensors, which form wireless sensor networks (WSN). These WSNs play a huge role in the way patients are being monitored. The wireless sensor networks generate a huge amount of data. To explore these massive amounts of data that are generated, there is a need for different kinds of IoT services, such as computation resources, storage capabilities, distribution system, high processing, and others. These networks of connected sensors send the generated data to central storing and computing devices, called the Cloud. The data generated from a network of sensors are to be sent to the cloud, as the cloud infrastructure is capable of performing tasks of analysis, aggregation, and storage. But this simple sensor-to-cloud architecture, where all the generated data are directly moved to the cloud and rely on the cloud for various services, such as storage, are analyzed to determine decisions regarding various actions that are not that viable.

The reasons why sensor-to-cloud architecture is not viable are as follows:

1. When the cloud is relatively far from the sensor generating patients' data, it becomes a problem for the latency-sensitive applications. In healthcare, there are many emergency response systems that require real-time operations, in which efficiency and time play an important part. There is high latency due to delay caused by the cloud. As there is delay in the transfer of data from the sensors to the cloud, due to many factors such as internet connectivity and bandwidth issues, and after the data reaches a distant cloud server, the data is analyzed and then the information is sent to the hospitals or personal physicians. In this scenario, the healthcare system couldn't match industrial control systems, which require a low-delay response time. Hence, the transfer of such immense amounts of data back and forth is not an efficient option, not only due to latency issues, but also due to security.
2. With the development of healthcare IoT systems, the amount of data and data nodes (such as sensors or medical devices) keep increasing, resulting in automatically increasing the burden on the cloud, and there is also an issue of network bandwidth, as huge amounts of data at one go is transferred via the internet to the cloud, and most hospitals would not prefer patient data to be stored outside. Also, there is always the possible case of there being a network failure or a cloud center failure, which puts patients' health at risk. Table 11.1 summarizes the various differences between cloud and fog computing.

The new architecture is introduced, where the classical centralized cloud computing architecture was extended to a distributed one, which provided a quick reaction to the underlying medical device, and can reduce the burden and dependency on the cloud.

A distributed architecture refers to tasks being divided and then offloaded to more than one node and also acts as an intermediate device to store and handle massive amounts of data generated by the IoT medical devices. A new infrastructure paradigm called fog computing is introduced, which is a new computing concept, to tackle the limitations of cloud computing. According to Cisco [9], fog computing

TABLE 11.1
Comparison Between Cloud and Fog Computing

Items	Cloud Computing	Fog Computing
Latency	High	Low
Hardware	Scalable storage and computing power	Limited storage and computing power
Location of server nodes	Within the internet	At the edge of the local network
Distance between client and server	Multiple hops	One hop
Working environment	Warehouse-size building with air conditioning systems	Outdoors (e.g., streets, gardens) or indoors (e.g., restaurants)
Security measures	Defined	Hard to define
Attack on data	Less probability	High probability
Deployment	Centralized	Distributed
Location Awareness	No	Yes

is a part of the cloud computing paradigm that takes the cloud closer to the edge of the network. It provides a highly virtualized model of computation, storage, and networking resources between end devices, and classical cloud servers [10].

Yi et al. [11] have provided a general definition of fog computing. It is stated as:

> Fog Computing is a geographically distributed computing architecture with a resource pool which consists of one or more ubiquitously connected heterogeneous devices (including edge devices) at the edge of network and not exclusively seamlessly backed by Cloud services, to collaboratively provide elastic computation, storage and communication (and many other new services and tasks) in isolated environments to a large scale of clients in proximity. [11]

The main idea of fog computing is to relocate some data centers' tasks to fog nodes on the edge of the server. Thus, it has the advantage of a high data transfer rate, thereby causing a drastic reduction in user response time.

Fog computing is regarded as a novel architecture, which is built on the edge servers, that provides the limited computing, storing, and networking services in a distributed manner between end devices and the cloud. Fog computing is implemented at three networking levels:

1. The collection of data at the sensor level.
2. Then collected data is sent to an intermediate device or node, where data is processed in less than a second, along with decision-making to provide logical intelligence to sensors.
3. The intermediate device or node filters the data for the cloud and sends required data in a timely interval for the purpose of implementing large storage and further rich computing analysis.

The fog computing, as a new architecture, implies a number of characteristics that makes the fog a significant extension of the cloud, such as [11, 36]:

- **Adjoining physical location:** The fog is placed much closer to the medical devices deployed locally as the neighboring sink point than the cloud, which is a floating sink, deployed at a distant place from a medical device generating data. Due to the deployment of fog nodes at the network edge, it reduces time delay, and these end devices (sensors), if located in the same local area network, reduce transmission delay as compared to wide area networks. Fog computing supports location awareness in which fog nodes can be deployed in different locations.
- **Support for various communications networks:** The sensors and end devices support a wide range of communication networks, which becomes convenient to communicate data easily. Some of the protocols are supported by the fog, such as ZigBee, WiFi, 2G/3G/4G, WiMax, 6Lowpan, and so on. However, the network supported by the cloud is the single one, which is the TCP/IP protocol.
- **Service is provided by smart but not powerful devices:** The fog nodes are driven by smart devices such as personal computers or personal mobile phones, tablets or gateways, routers, or other cheap devices, where they perform limited computing for short and fast response. The work of data mining and long-term analysis is assigned to the distant cloud, which is located at a physical point driven by the costly and powerful servers.
- **Real-time interactions:** Fog computing applications provide real-time interactions between fog nodes rather than the batch processing engaged in the cloud.
- **Distributed computing:** Fog computing works on distributing rather than in a centralized manner. The fog is a distributed model and can be deployed anywhere. For any huge computation efforts, the layer is decomposed into smaller parts and assigned to many fog nodes, which are part of the layer for processing. These fog layers are distributed across the network. Due to the increase in amount of data and heterogeneity of data, network bandwidth becomes a major issue when a large amount of data is uploaded from sensors to the cloud at one go. A fog node in fog computing is responsible for dealing with the regional data, and the distributed computing strategy helps to reduce the computational complexity. Finally, the computed data by the fog nodes are continuously uploaded to the cloud, which is beneficial to solve the bandwidth issue.
- **Interoperability:** Fog components can interoperate and work with different domains and across different service providers.
- **Support for online analytics:** The fog can provide a real-time analysis and give the appropriate response actions to the end devices as the fog stores and computes the data from the end devices.

The fog computing expands the capabilities of the cloud near to the end user and provides storage, computation, and communication to edge devices. Fog computing brings many benefits for IoT devices [9, 12]. These benefits can be summarized as follows:

- **Low and predictable latency with a real-time response:** As fog computing works close to the device sensor, it helps to reduce data transmission overheads and does local computation on data. And it is able to respond very fast to the data.
- **Greater business agility:** With the use of the right tools, fog computing applications can be quickly developed and deployed. In addition, these applications can program the machine to work according to the customer's needs [9].
- **Privacy:** The confidential part of the data is filtered, if it is not required for the storage in the cloud, and the fog node processes the data locally before sharing it to third-party servers. This considerably improves the privacy and security of medical data.
- **Recover against cloud/network failure:** In case of network or cloud failure, the fog layer enables the safe recovery of application and data. The fog has transient storage where the data from sensors can be stored, and these data can be stored when needed.
- **Scalability, redundancy, and elasticity:** One fog node or many computation nodes that are connected jointly can be used to build fog computing infrastructure, and when more computing is required, it is possible to add more fog nodes, which significantly improves scalability, redundancy, and elasticity.
- Also, due to the fact that data is stored in data centers in the cloud, the problem of infrastructure, maintenance, upgrades, and costs is solved.

Fog computing is considered the best method for designing healthcare applications because these applications need latency sensitivity, show low response time, and produce a large amount of data.

11.2 ARCHITECTURE AND THE ROLE OF FOG COMPUTING IN IOT HEALTHCARE SYSTEM

The fog computing can be used to have faster devices, which is very important in real-time operation. Maximum healthcare devices are used for real-time applications.

11.2.1 IoT e-Health Ecosystem Architecture

With the previous e-Health ecosystem, which had a straightforward design approach where data generated from medical devices (e.g., ECG machines), which are not capable of storing their generated data, were transferred to a cloud for processing. As a large number of medical connected devices, the latency of the connection with

TABLE 11.2
Layered Architecture of Fog Computing (Top to Bottom)

Physical and Virtualization Layer	Virtual Sensor and Virtual Sensor Network
	Things and physical sensors, wireless sensor networks.
Monitoring Layer	Activities monitoring, power monitoring, resources monitoring, response monitoring, and service monitoring.
Pre-Processing Layer	Data analysis and data filtering, reconstruction and trimming.
Temporary Storage Layer	Data distribution, replication, and deduplication. Storage space virtualization and storage devices (NAS, FC, I SCI, etc.).
Security Layer	Encryption/description, privacy, and integrity measures.
Transport Layer	Uploading pre-processed and secured data to the cloud.

the cloud was significant. Moreover, these devices are power- and bandwidth-constrained, which makes them unfit directly to the cloud architecture. With the coming of fog computing, which is an important paradigm transfer toward a hierarchical system architecture. Fog computing is highly virtualized and provides a medium for computing, storage, and networking between end devices and the cloud. Fog is an intermediate computing layer between the cloud and end devices that complements the advantages of cloud computing by providing additional services for the emerging requirements in the field of IoT. The main notion of fog computing is to migrate the tasks of data centers to fog nodes situated at the edge of the network. The internal layered architecture of fog computing consists of six layers—physical and virtualization, monitoring, pre-processing, temporary storage, security and transport layer—as shown in Table 11.2 [13, 14]. Here a holistic multi-layer IoT e-Health ecosystem/architecture is discussed. The system architecture includes the following main three components, as shown in Figure 11.1.

1. IoT e-Health Device Layer.
2. IoT e-Health Fog Layer.
3. IoT e-Health Cloud Layer.

11.2.1.1 IoT e-Health Device Layer

In this layer, there is a huge set of IoT-based medical devices, such as wearables, sensors, heart rate monitor, glucose monitor, blood pressure monitor, and body temperature monitor, etc., which are enabled with omnipresent identification, sensing, and biomedical and context signals. These medical devices capture data from the body and the room at any given time, and this data, which is coined as the health status data of the patients. The health status data is synced securely by any personal

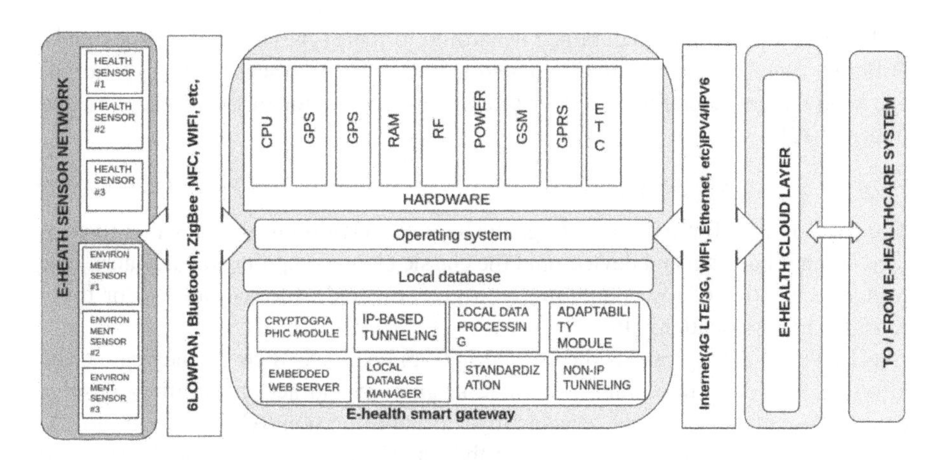

FIGURE 11.1 Internal architecture of the fog layer.

computer or cell phone in real-time, and later the data is sent to the e-Health cloud platform [15]. Along with health status data, the context information [37] (e.g., date, time, location, temperature) is also communicated to fog nodes. Context-awareness enables one to identify abnormal patterns and make more precise inferences about the situation. Other sensors and actuators (e.g., medical equipment) can be also connected to the systems to transmit data to medical staff, such as high-resolution images (e.g., CAT scan, magnetic resonance imaging).

In the first layer, different wearables or sensors, such as a smart-watch, or smart eyeglasses, are used to acquire the patients' data. All they require is a connection using a suitable communication protocol via fog node. In this specific scenario, a tremendous amount of personal area network (PAN) protocols and wireless sensor network (WSN) protocols are used for proper connection, such as WiFi for remote connection and Bluetooth Low Energy (BLE) for short distance connection. Further, IoT-based MDs can be divided into two groups:

1. Physical smart sensors are used to digitally monitor patients' health and track their physical wellness. These are used to monitor things like ECG, body temperature, heart rate, hemoglobin, glucose monitor, blood pressure, and pulse monitor.
2. Virtual smart sensors are used to capture the patients' health information and contextual data from the surroundings. The virtual smart sensor provides remote monitoring, diagnostic, remote consultation, and patient health records. These sensors are highly reliable, and low cost. Therefore, they can be broadly distributed at various public hospitals or clinics to monitor the patients' health status continuously. These geospatially distributed sensors generate enormous sensing data streams, which have to be processed carefully.

As shown in Table 11.2, the physical and virtualization layer involves different types of nodes, such as physical nodes, virtual nodes, and virtual sensor networks. These nodes are managed and maintained according to their types and service demands. Different types of sensors are distributed geographically to sense the surroundings and send the collected data to upper layers via gateways for further processing and filtering [12].

11.2.1.2 IoT e-Health Fog Layer

The fog provides limited computing, storing and networking services in a distributed manner between end devices and the classic cloud computing data centers. The primary objective of fog computing is to give low and expected latency for time-sensitive IoT applications [16].

e-Health is among one of the critical IoT applications where latency cannot be afforded. It is very important to be able to analyze and act on the time-sensitive data and circumstances, such as acute myocardial infarction (AMI), or heart attack, in a few seconds. Therefore, relying on the traditional cloud model architecture to collect and analyze the patients' sensitive medical data, which is located across a wide distant geographical area in the presence of various environmental conditions, is problematic. It takes more time to make decision-making of time-sensitive data, where low latency is required. To overcome this issue, the most efficient approach is to introduce fog or edge computing to take down the cloud computing high-latency problem. The raw data is generated from IoT medical devices, and is forwarded to fog layers, where the data is processed and forwarded for further analysis. The fog layer is equipped with various low-power and high-performance computing nodes and has a small storage space called fog nodes. These fog nodes are connected to many medical devices, and these fog nodes are highly capable to maintain reliable and secure connectivity between IoT medical devices. It provides the required intelligence to the end devices and filters the data for the cloud, where the primary objective is to have low and predictable latency in the latency-sensitive IoT applications. The fog layer periodically sends the aggregated data to the cloud for long-term storage and big data analytics. It also receives the commands as well as the configuration data from the cloud. Since fog nodes might not have a valid public IP, it is important to ensure the reachability of the fog nodes from the cloud [39]. To achieve this goal, different mechanisms are used, such as WebSocket, MQTT, and IP tunneling. The fog layer also has its own storage space for keeping the critical data where critical data are analyzed. Fog nodes provide a bi-directional flow of data. This data, which is routed to the cloud from IoT medical devices to the fog layer, will be available to doctors or patients on request. This also solves the network bandwidth issue, as a lot of raw generated data and for nodes perform compressing, filtering, aggregation, and formatting of raw medical data collected from IoT medical devices, and the required data is sent to the cloud. These nodes are placed closest to the medical devices that produce the data. The fog layer also provides multi-layer security measures for encryption, authentication, and access control to protect the patients' data and privacy. Fog nodes have flexible integration. It can be interoperable with this wide range of medical devices and wide range of interfaces, and the fog nodes are equipped with multi-standard interfaces. Popular communication protocols that can be used in this layer are Wi-Fi, 3G/4G, radio frequency identification (RFID), BLE, Zigbee, and wired protocols such as Ethernet, etc.

While at the monitoring layer, resource utilization and the availability of sensors and fog nodes and network elements are monitored. All tasks performed by nodes are monitored in this layer, monitoring which node is performing what task, at what time, and what will be required from it next. The performance and status of all applications and services deployed on the infrastructure are monitored.

For example, an ECG device can generate a huge amount of raw time-series data within a day.

However, it is not feasible–and not even necessary–to send these huge amounts of data from thousands of patients to the cloud. And these will be a network bandwidth issue here, as a huge amount of data is sent via the internet to the cloud. But these fog nodes are reasonable to process, filter, and compress the data traveling between the medical devices and the cloud.

11.2.1.2.1 Properties and Features at the Fog Layer

11.2.1.2.1.1 Local Data Processing The data from IoT medical devices is processed at the fog nodes, and it is implemented to provide intelligence at the fog layer, which also acts as a gateway by which the streaming data is analyzed locally. This includes data mining and pattern recognition in time-series data, feature extraction, machine learning, light-weight signal processing, event processing, automated reasoning, embedded web server, and embedded WebSocket server, etc.

For medical cases, where the system needs to react as fast as possible in medical emergencies [38], the fog layer continuously handles a large amount of sensory data in a short time and responds appropriately with respect to various conditions.

Different local data processing that takes place at the fog layer:

- **Data filtering**: It is essential to implement appropriate pre-processing at the edge after receiving data from various sensors before proceeding with any advance processing, such as data analysis. Usually there are noises and outliers in the data, which arrive from the sensor. The fog layer addresses this issue, as it has an interface with the sensors directly. The fog layer receives digitized signals from sensors via various communication protocols. In Table 11.2, the pre-processing layer performs data management tasks. Collected data are analyzed and data filtering and trimming are carried out in this layer to extract meaningful information. Different medical devices, like ECG, EEG, and EMG, generate bio-signals, which are collected from users' bodies for assessing a patient's health status. During the sensing from the human body, noise also accumulates to the bio-signals and distorts the signal quality. These noises are filtered out in the fog layers.
- **Data compression**: Data compression is used in the context of data communication for lowering communication latency and energy consumed during a transaction. In the eHealth IoT system, both lossy and lossless compression methods are useful. In the case of resource-constrained sensors, due to limitations such as battery lifetime and available processing power, lossy data compression is more suitable.

- **Data fusion:** Data fusion enables the system to effectively decrease the volume of data by which the energy needed for data transmission is reduced. Data fusion is classified into three classes [40]:
  - **Complementary data fusion**: This is done to obtain a greater global understanding and knowledge, e.g., obtaining a difference in temperature between the body and the surrounding environment is a specimen provided from two sensory data.
  - **Competitive data fusion**: This is done to improve the accuracy and consistency in case of sensor failure at edge level. Here a single parameter is collected from different sources.
  - **Cooperative data fusion:** This provides benefits at the edge, and here new information is extracted in smart gateways from the diversified data collected from diverse sources. For example, cooperative data fusion provides complete information about the medical state of a patient from his/her vital signs.
- **Data analysis:** The local level analysis at edge improves the sensitivity of the system and the ability to detect and predict emergency situations and reacts to the emergency situation faster. By utilizing data analysis in the fog layer, it enables the system to minimize the processing latencies of critical parameters. Local data analysis and local feedback improves the system reliability and consistency in case of the unavailability of an internet connection. The fog layer keeps the data from IoT medical device data locally and processes it and stores the analysis locally and synchronizes them with the cloud later.
- **Local storage:** For health systems, smooth recovery of data fog layers stores the approaching data from sensors or devices in a local storage. The operating system at the fog layer is responsible for handling the local repository and stores the data in a non-volatile memory, as shown in Figure 11.1. Based on the importance of stored data, the information is kept in compressed or encrypted format [41]. It makes the system reliable and robust even when the network is unavailable. Local storage also acts as a cache to implement a continuous data flow when the speed to transfer data from the fog nodes to the cloud is limited by network bandwidth. The pre-processed data is then stored temporarily in the temporary storage layer in Table 11.2. When the data are transmitted to the cloud, they no longer need to be stored locally and may be removed from the temporary storage media [14, 17].
- **Local actuation**: An actuator works in the reverse direction of a sensor. It takes an electrical input and turns it into physical action. Actuation can be used in the form of information streaming, sensor network reconfiguration, and controlling medical actuators.

An expected and fast response time is demanded from the actuator. Fast-response actuations can be an adjustment of frequency of electrical nerve stimulation based on heart rate. Notifications are mandatory features for smart e-Health gateways at the edge of the network. In emergency situations, health monitoring systems inform and notify medical teams, caregivers, and the patients. Notification service failure can cause serious problems for both patients and medical treatments. In comparison to a cloud server, which is able to send notifications via several methods, a gateway has limited resources and can only notify via some certain medium. However, the advantage of gateway-based notifications is that it acts independently (e.g., via the local network or GSM) even during the unavailability of a cloud server, which makes the system more trustworthy and ensures that users can receive critical notifications in time.

- **Security and data protection**: To protect the patient's data, there are multi-layer security measures for authentication, and encryption at the fog layer. In Table 11.2, in the security layer, the encryption/decryption of data comes into play. In addition, integrity measures are applied to the data to keep them safe from tampering. A high level of security can be provided by the operating-system level methods at gateways, such as IPtable provided by Linux. To be more exact, IPtables and IPFW are provided by Linux kernel firewalls, which are used for configuring IP packet tables. Typically, IP tables are configured to allow some ports for communication, while other ports are kept blocked for stopping unwanted traffic [43]. The gateway communicates over secure HTTPS when it acts as an embedded web server during network unavailability and authenticates sensor nodes to maintain the confidentiality, integrity, and authenticity of the system. For data protection, cryptographic operations on data are required. But cryptographic operations need heavy processing power and energy, making them unfeasible for resource-constrained devices. In a proposal focused on security, Rahimi et al. [44, 45] introduced a secure and efficient authentication and authorization approach for Health-IoT systems, which requires some processing power at the edge.
- **Protocol translation**: The standardization and interoperability plays a key role to the success of eHealth-IoT systems. With such a heterogeneous mix of networking technologies, protocols, and platform choices to implement IoT-based systems, integrating these applications is challenging. Our smart e-Health gateway plays an important role in providing interoperability for the various sensors connected via different definite network interfaces. The health sensors and the context sensors communicate the data to the Smart e-Health Gateway, using either wireless network or wired connections, while using different standards (e.g., ZigBee, 6LoWPAN, Bluetooth, Wi-Fi). The smartness of the gateway comes here in the form of easy integration of these heterogeneous networking technologies, protocols, and standards, and thereby it is possible to exchange information and work seamlessly.
- **Latency**: For remote monitoring of acute diseases and emergencies, a health monitoring system should be capable of making rapid decisions and providing quicker responses, for this data processing and transmission time should be minimal. In cloud computing, where raw data is transferred from sensor nodes to cloud, if the network condition is bad, it may cause uncertainty to response latencies. The situation becomes worse when streaming-based data processing is needed (e.g., signal processing on ECG or EEG

signals). By implementing high-priority data analytics in distributed smart gateways, which helps in making critical and time-sensitive decisions within the local network, makes the system more robust. The processed data is later transmitted to the cloud for storage and further analysis according to network availability.

- **Energy efficiency for sensor nodes**: Due to resource constraints at sensor nodes, data processing has many drawbacks. In some cases, to execute complex algorithms successfully, a high cost of energy is required. Therefore, selective signals from sensor nodes are transferred to smart gateways at the fog layer for processing tasks. It is an effective solution to address the different issues, mainly when gateways are not battery-powered. Several approaches [46, 47a, 47b] have focused on providing energy efficiency for Health-IoT applications. Although such techniques enhance the energy efficiency of sensor nodes, a considerable amount of energy can be still saved via fog computing by transferring some loads from sensor nodes to smart gateways.

11.2.1.3 IoT e-Health Cloud Layer

In Table 11.2, finally, in the transport layer, the pre-processed data is uploaded to the cloud to allow the cloud to extract and create more useful services. It is the top layer of the architecture which contains a high distributed computing data center and storage capacity that implements data warehouse and data analytics of health data, which gives a long-term, complex, and behavior analysis, relationship modeling, and long-term pattern recognition. It also provides centralized controlling, demonstrations for web clients as a graphical user interface for final visualization, feedback, and city-wide monitoring. The health data and context data which are collected constitute a source of big data for statistical and epidemiological medical research. The cloud server also provides an interface for the patients and the healthcare professionals (e.g., medical doctors) to monitor and manage the connected IoT devices. It also enables the users to read or share their medical data [47].

It consists of different internal layers [15]:

- **Connectivity layer:** This layer is provided with a wide variety of built-in capabilities to establish connectivity between e-Health IoT devices, fog nodes, and the IoT e-Health cloud layer. In a fog, only selective localized applications are provisioned and synchronized with the cloud [33]. With the dual functions of the cloud, the data delivery and update from the cloud to fog faces problems related to communications sessions created during the processing of fog nodes. Picking the correct communication between the fog and the cloud that ensures high performance and low latency of fog nodes is a key challenge [18, 34]. The fog nodes can be connected to the cloud using any hardware over any feasible communication channel, like wired networks, wireless networks, 3G/4G/5G, based on a wide-range of protocols available like MQTT, WebSocket, REST API, etc.

- **Data management layer:** This layer integrates the cloud data from multiple sources. It captures the data from many fog nodes and stores the data safely and securely. Furthermore, data is integrated flawlessly with non-sensor sources, such as data from e-prescription websites, EHRs, and web sources. Consequently, patients and medical practitioners can access the data anytime, anywhere as and when required. This leads to significant increases in collaboration across all disciplines, increasing the efficiency of the healthcare plan. Another advantage of the cloud platform is that it separates the data layer from the application layer, while providing a unified schema in terms of capture and query transactions. This feature results in more flexibility to develop new applications. This module includes built-in capabilities for managing users, groups, devices and fog nodes, access permissions, and roles.
- **Application layer**: Different services provided by the application layer are as follows:
  - **Dashboard**: A web-based application that allows patients and those engaged in the patient's care to monitor and manage the connected e-Health devices, and the aggregated medical big data. The dashboard enables the users to remotely monitor, configure, control, diagnose, and repair the e-Health devices connected to the cloud. It also allows the users to view, visualize, understand, and share their data.
  - **Rule engine**: A comprehensive engine with a web-based UI, which is used for analyzing the incoming data and to create appropriate events, alarms, notifications, and trigger actions.
  - **Big data analytics**: It analyzes the aggregated medical data to effectively identify actionable insights. Here several event-processing, rule-based processing, data-mining, machine learning, and reasoning-based algorithms can be applied on the stored historical records to extract meaningful insights about patients' health.

For example, this analysis about the historical data might be helpful in predicting the birth of a new disease, so that the patient can take early precaution measures, which reduce the cost and help in saving the patient's life.

Communication between layers: Fog computing combines to form a single platform to provide various applications and services at the edge of the network. At the edge, the data generated by eHealth IoT devices and sensors are communicated to fog nodes, and there data is processed by fog nodes as soon as they receive the data, and then transfers it to the cloud for further processing and analysis. The cloud connects to the fog layer after receiving the data from the fog layer, stores the data and performs data mining, and responds with an analysis back to the fog nodes. This analysis is used by doctors to monitor the patients' health. The fog layer works as a mediator between the IoT eHealth device layer and the eHealth cloud layer. The data generated by the medical device layer is uploaded to the fog layer. Then, the fog layer ingests this data and processes it.

Then, the fog layer updates data to the cloud layer for deep analysis and for long-run storage. In short, fog layers work as a pushing service to both layers. For efficient power utilization, only a portion of collected data is uploaded to the cloud. In other words, the gateway device connecting the IoT to the cloud processes the data before sending them to the cloud. This type of gateway is called a smart gateway [19]. Data collected from sensor networks and IoT devices are transferred through smart gateways to the cloud. The data received by the cloud is then stored and used to create services for users [14]. Based on the limited resources of the fog, a communication protocol for fog computing needs to be efficient, lightweight, and customizable.

- **Switching network**: Different sensors produce different format data, and there are different formats requiring different communication protocols to transfer it in between layers. At that moment, the most important thing the fog computing does is to switch networks among cloud and IoT devices by establishing the communication link in between layers. Fog computing requires supporting different network protocols, for example, WiMax, ZigBee, WiFi, 2G/3G/4G/5G. But only the TCP/IP protocol is supported by the cloud.
- **Aggregation and filtering**: Once the medical devices and sensors generate data, there could be a probability that after some period, it is no longer needed to upload data to the cloud. For handling such situations, the fog layer has additional functionality for data processing, such as aggregation and filtering. Aggregation is done on data received from a medical device before uploading to the cloud. The fog layer is capable of deciding what content to send, when to send, that is, time and how to send, and that is the format of data. In this process, fog layers remove a few invalid and duplicate data. Further, unnecessary and complementary data is aggregated in time and space dimensions. For this, there is a need for online analysis on real data with some standardization. Moreover, the fog computing architecture supports transient and semi-permanent storage. With this transient storage, data has to be optimally stored with limited storage size. Fog layer computes on real-time data such as e-Health care data to apprehend online analytics intelligence. Fog layer-based intelligent computing is a complex job, as various applications have different calculation parameters. [20]

As shown in Figure 11.2, the working of the architecture, where data of different variety, velocity, and volume are generated at the sensor end and are transferred to nearby fog nodes. Here the data are processed and analyzed, and if any alarming situation is found, alert notifications are pushed to caretakers, doctors, and patients. Otherwise, only the required data are transferred to the cloud layer for permanent storage and analysis. These analyzed patients' data can be accessed by the doctors, hospitals, and patients through various interfaces, such as mobile applications or web.

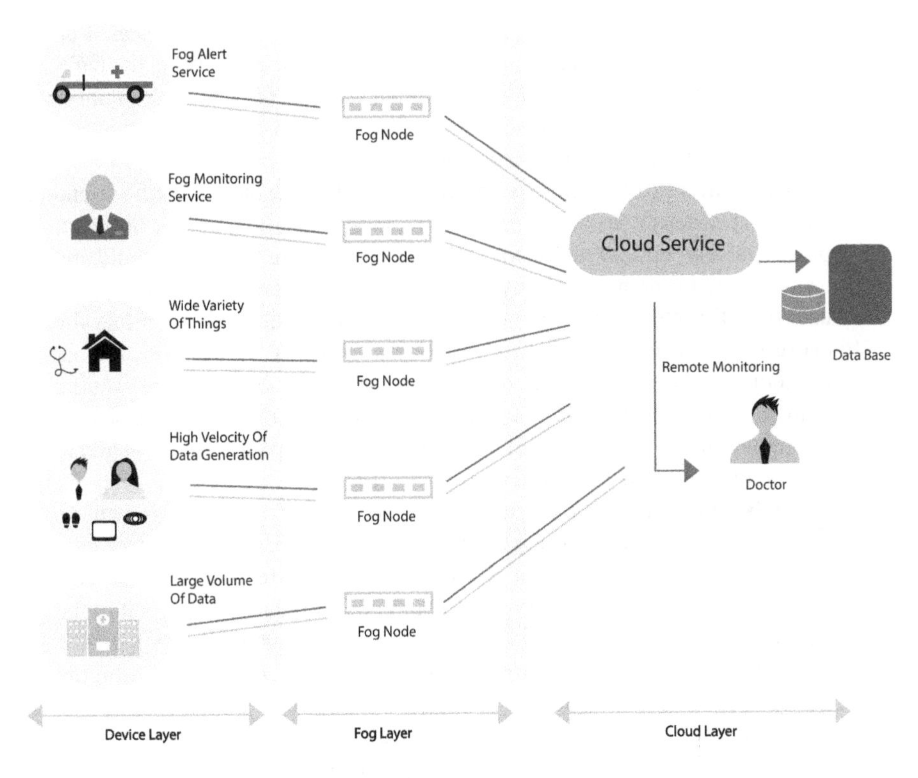

FIGURE 11.2 Overall flow of the architecture.

11.2.2 IoT e-Health Challenges

There are several research challenges that the IoT e-Health system faces:

1. **Data management**: IoT e-Health systems have to handle the complexity of huge amounts of patient data, which faces big data 5 Vs' data value, volume, veracity, variety, and velocity [21–23]. The challenges due to data variety come as the fog layer receives data of different formats, depending upon the healthcare application, and different health applications send data in different forms, e.g. ECG data is sent in XML format, while skin diseases using camera-based IoT devices need to handle image formats [24]. Then the challenges of data volume and velocity are more related with the capabilities of fog node hardware to receive, process, store, and communicate the high-resolution data received from medical devices that could be with patients or in hospitals or clinics. With increasing complexity of data, the fog node faces major challenges in receiving, storing, and communicating the data.
2. **Scalability:** The data is collected from a large number of medical sensors, which generates a huge amount of data and a huge amount of resources such as data storage, and can be used for further processing and analysis. The real challenge is to be capable of responding to the rapid growth of

IoT devices and applications [18, 25]. There should be a proper scaled-up facility to store such a huge amount of data, and at the same time, the facility that the entire hospital or a patient in the hospital or remote at a remote place can access the medical services, such as checking updates and health status monitoring by their smartphones.

3. **Interfaces and human-factors engineering**: The interface of the front-end technologies sensors is one of the key factors, such as the interface of smartphone, tablets, etc. As these IoT medical devices are high-tech tools and users or patients do not have any experience of using them, one has to create user-friendly interfaces and patient-centric medical devices, such that the end-user could self-train themselves. End-users don't have any idea about the setup of wireless network, devices, and sensors operations and syncing of medical devices. So setting up of these devices should be straightforward and autonomous. Hence, device interfaces need to be patient-friendly and to require minimum involvement of experts [26].
4. **Security and privacy**: Security and privacy of IoT e-Health systems is present across the whole lifecycle of the system, starting from sensor data generation, to implementation and deployment. There is a potential risk that every single IoT medical device could be exploited to either injure the end-users or to risk the privacy of the end-user, which can also risk patients' safety [27]. However, protecting the IoT e-Health ecosystem is a highly complex and challenging task, where the key concern is that end-users' data should not be acquired by unauthorized agencies, which creates risks to personal safety [28–49]. A multi-layer approach to overcome the challenges and protect the IoT medical device is discussed below.
   - **Device layer**: A medical sensor which captures the real-time data and is connected to medical devices, gateways, fog nodes, and mobile devices that capture, aggregate, process, and transfer the medical data to the cloud. Although fog nodes will need to be protected by using the same policy, controls, and procedures, and they use the same physical security and cybersecurity solutions [9], the fog environment itself is vulnerable and less secure than cloud computing. The most common attacks at the device layer are tag cloning, spoofing, RF jamming, cloud polling, and direct connection. In cloud polling, attackers inject commands [28, 29] to devices which try to redirect the network traffic. This is carried out by many means, such as Man-in-the-Middle (MITM) [47] attack and changing the settings of domain name system (DNS) settings. To overcome such situations, IoT devices should evaluate and verify certifications received that really belong to the e-Health cloud. In the direct connection attack, attackers can use Service Discovery Protocol, such as Universal Plug protocol, to find and discover IoT devices. To overcome this type of attacks, IoT e-Health devices should ignore, reject, and block the unauthenticated requests. Overall security measures at the device layer include identity, authentication, and authorization management, preventing unauthorized applications to be executed, fine-grained access control capability of resources, whitelisting application

sandboxing, protection of data during capture, storage, and traffic filtering feature, transit, secure pairing protocols, password enforcement policies, fault tolerance, and secure transmission mechanisms [30]. But implementation of these security algorithms in IoT e-Health devices are quite challenging and should be carefully implemented, as these devices are low-powered and have extremely limited resources.

- **Network layer:** This layer manages to create proper and suitable connections between sensors, IoT e-Health devices, fog nodes and e-Health cloud, based on different types of network protocols (such as Wi-Fi, BLE, ZigBee). The most common attacks at this layer are Sybil attack, Eavesdropping attack, Man-in-the-Middle attack, Sleep Deprivation attack, and Sinkhole attack. Many research studies focus on cryptography and authentication to improve network security to protect against cyber-attacks in fog computing [11, 18]. In order to secure the network layer, the use of trusted routing mechanisms, such as message integrity verification techniques (using hashing mechanisms such as MD5 and SHA), as well as point-to-point encryption techniques based on cryptographic algorithms, is very important. Mostly cryptographic algorithms like symmetric algorithms (AES, DES, Blowfish, and Skipjack) are used as symmetric algorithms and are less compute-intensive, and are more suitable for low-power 8-bit/16-bit IoT devices [31].
- **Cloud layer**: In this layer, mostly the companies that deliver e-Health products/services need to have a capable and adequate mechanism to encounter the adverse impact of attacks. The top common cloud susceptibilities are Denial-of-Service (DoS) attack, sniffing attack, Trojan horses, Spear-Phishing attack, SQL injection, malicious code injection, and brute-force attack (using weak password recovery methods) [29].

5. **Energy consumption**: The fog environment has a huge number of fog end devices. The computation in centralized cloud models is more energy-efficient than the computation of distributed cloud models. Therefore, reducing energy consumed in fog computing is an important challenge that needs to be addressed [32].

11.3 CASE STUDY

The following case study is an illustration of the role of fog computing in an automated EWS health monitoring system, which intelligently monitors vital signs using Internet-of-Things (IoT) technologies.

11.3.1 Health Monitoring System Using Early Warning Score: Case Study Based on Internet of Things Using Fog Computing

Early warning score (EWS) is an approach to discover the declining health of a patient.

It is based on the fact that there are several changes in the normal functions of human organisms and their parts prior to the clinical deterioration of a patient.

Remote patient monitoring is crucial for a large number of patients that are undergoing acute diseases, such as different heart conditions. Traditional EWS procedure was mostly used for in-hospital clinical cases and was performed in a manual paper-based fashion. To predict or early-detect future potentially critical circumstances, proper medical services are required that consider the ongoing medical state of the patient, and this could be achieved by continuous health monitoring of patients. Early detection of health deterioration can be a key aspect to reduce the mortality rate and healthcare expenses related to several diseases. IoT-enabled health monitoring systems record physiological parameters and vital signs from a wireless body area network (WBAN), which is a set of medical sensors attached to a patient's body [50], and sends them to a cloud server for further processing and storage. In order to predict health deterioration, a technique called Early Warning Score (EWS) [51] has been proposed. In this system, medical sensors record the patients' vital signs in an observation chart at certain time intervals and assign a score to the value of each sign in its range. A higher score means more abnormalities of a specific vital sign, and the sum of all scores indicates the overall health status of a patient [52]. The overall patient score, which is the sum of all individual scores, is then used to decide whether the patient is deteriorating or not [53]. The Early Warning System is a methodology for early detection of health deterioration to minimize the impact of sudden severe changes in health. Such a system uses a process called Early Warning Score (EWS) to calculate different scores from a patient's observation chart. Table 11.3 shows a typical EWS model where scores for different physiological parameters are shown. These measurements are based on repetitive physiological measurement of vital signs to derive a composite score, which is used to identify whether a patient is at a risk of deterioration. Studies in this field have shown that patients often have signs of clinical deterioration up to 24 hours prior to a serious clinical case, requiring a full intervention [54–56].

TABLE 11.3
A Typical Early Warning Score Model [57]

Physiological Parameter	3	2	1	0	1	2	3
Respiration Rate	≤8		9–10	12–20		21–24	≥25
Temperature (°C)	≤35.0		35.1–36.0	36.1–38.0	38.1–39.0	≥39.1	
Heart rate (beats/min)	≤40	91–100	101–110	111–219			≥220
Oxygen Saturations (%)	≤91	92–93	93–95	≥96			
Any supplemental Oxygen		Yes		No			
Level of consciousness				A*			
Blood pressure (mm Hg)	≤90	91–100	101–110	51–90	91–110	111–130	≥131

TABLE 11.4
Risk Level Associated with Calculated Score

Score	Risk Level
0	Low
Aggregate 1–4	
Red Score	Medium
Aggregate 5–6	
Aggregate 7 or More	High

The simplest type of score can be calculated as level of consciousness, pulse rate, systolic blood pressure, respiration rate, body temperature, and blood oxygen saturation. Each parameter has a maximum score of 3 and a minimum score of 0 from which the final score can be calculated. Table 11.3 shows a typical early warning score model. After calculating the early warning score, the treatment order of the patient can be changed, based on the score results. Lower scores lead to change in observation frequency, and higher scores result in providing a higher level of medical care for the patient, such as assessment by a health expert or transferring to the intensive care unit. An automated EWS health monitoring system to intelligently monitor vital signs and raise notification alerts, depending upon the level of risk, as shown in Table 11.4, which prevents health deterioration for in-home patients using Internet-of-Things (IoT) e-Health monitoring system. As shown in Table 11.4, if the aggregate score lies between 1 and 4, then the risk level is low, which means the person is safe and normal. Similarly, for aggregate scores in the range between 5 and 6, the risk level is medium where the patient gets an alert about the health problems, and a score above 7 is considered as high risk, where the patient needs emergency care.

IoT can provide a real-time 24/7 service for health professionals to monitor from a distance the in-home patients via the internet and receive notifications in case of emergency. The new advance EWS system, where continuous reads, transfers, record, and process the vital signs, has been implemented.

An IoT health monitoring system's architecture is defined into three layers:

1. **Sensor network:** This layer is closest to the body of the patients and its work is to continuously collect patients' data.
2. **Smart gateway:** The gateway acts as a bridge and builds a continuous communication channel between sensors and cloud with various communication protocols, such as Bluetooth and Zigbee, etc.
3. **Cloud layer:** The cloud layer is a processing unit where data from sensors are received via gateway, stored for a long time, and analyzed.

Figure 11.3 shows the three-tier architecture of IoT health monitoring based systems for remote patient monitoring data. The system provides actions such as response, identifying, and recommendation to a patient. This type of architecture not only monitors the current state of the patient, but also can provide future prediction on

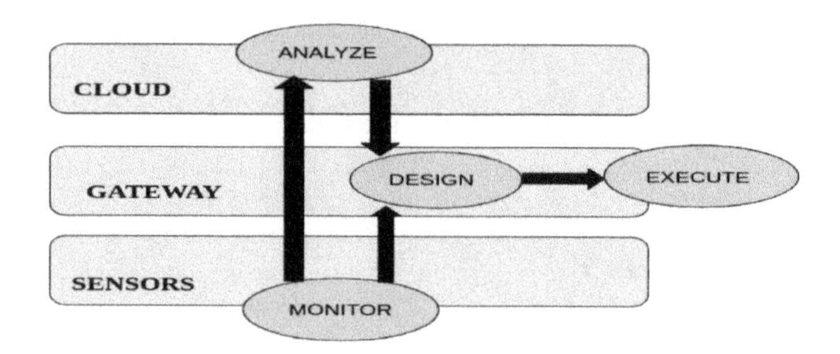

FIGURE 11.3 Three-layer fog based e-health architecture.

medical conditions that a patient may face. This is done by applying big data analytics and machine learning methods to patients' data at the cloud layer. These analyzed data can be used by many patients suffering from different kinds of disease and by doctors for monitoring and diagnostic processes. These can be assessed using different accessing platforms, like web dashboard or apps. The system might fail to perform in case of unavailability of internet connection and also due to network bandwidth issues. So to overcome such problems, gateways are provided with a considerable computational capacity, and which can process the data from the medical sensor before the cloud layer can do. By computing sections, we come closer to the sensor layer.

The system architecture of our IoT-based early warning system is shown in Figure 11.3. The system architecture includes three main components, as described in the following sections.

11.3.1.1 Sensor Network

At the first layer, which is known as a network of sensors, its primary task is data collection using different medical sensors. This layer is categorized into three groups, as shown in Figure 11.4, based on their data conversion rate and their function. The first and essential group includes medical sensors to scan important signs such as heart rate, blood pressure, respiration rate, blood oxygen level, body temperature, and ECG. The second group of sensors contains contextual sensors for humidity of the room, recording the light and temperature around the patient. The third and last type is activity sensors, which note the patterns of movements, posture, and total daily steps of the patient, and then the collected data is forwarded to a smart e-health gateway, as shown in Figure 11.5.

11.3.1.2 e-Health Gateway

The gateway can receive data from various types of medical sensors via a UDP server. This smart gateway stores data for each sensor in different files, giving consideration to the information of the patient whom the data is anticipated from, and provides some services such as data compression and storage in offline mode. As features of the collected data are not homogeneous, a tailoring is needed to unify the data to form a proper structure. The rate of data collection is also different for different sensors, which fluctuates from 250 samples per second (e.g., ECG) to

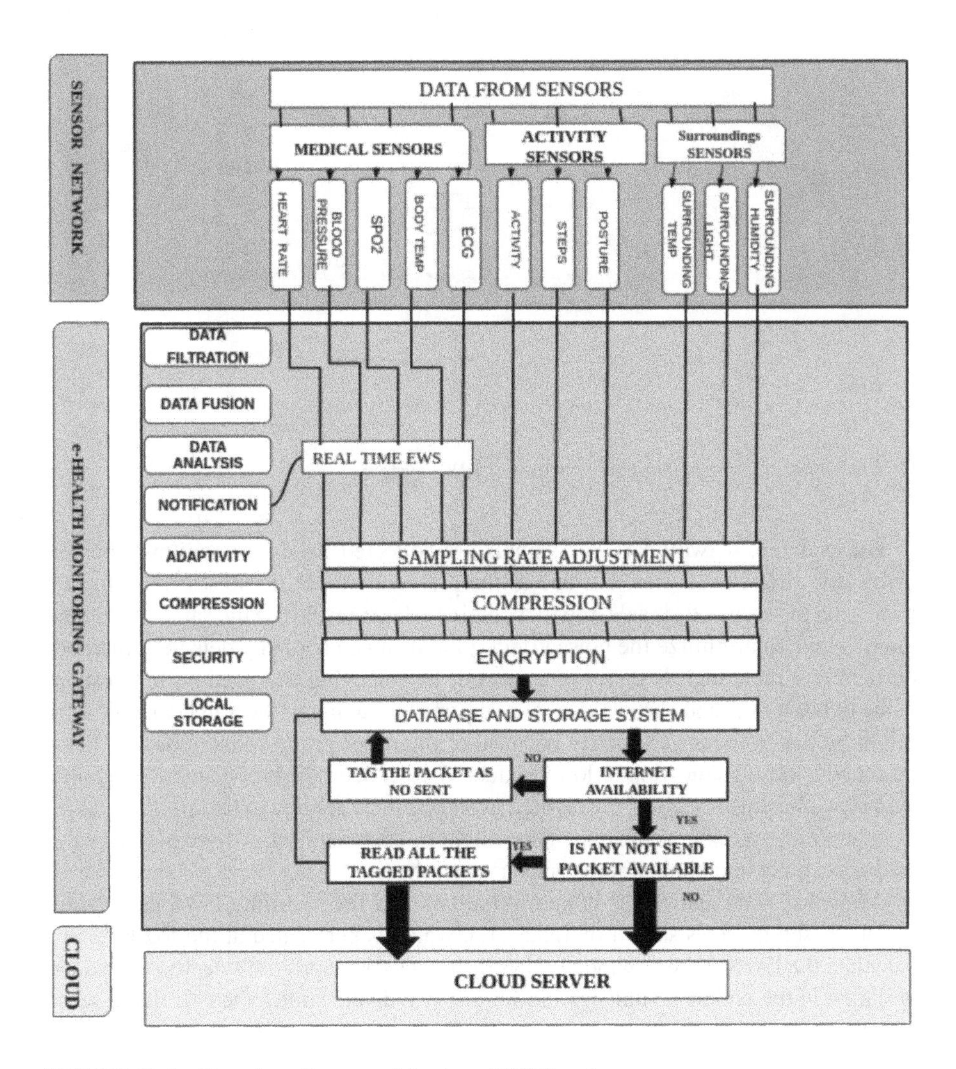

FIGURE 11.4 Data flow diagram of fog-based EWS system.

about 5 samples per day (e.g., step counts). So, different communication protocols are used for different types of sensor data transmission (e.g., Bluetooth and Wi-Fi with UDP and TCP transmission protocols). The data from different medical sensors have noise or unnecessary data. Noise from data is first filtered out at the smart gateway.

For example, a bandpass filter (0.5–100 Hz) with Finite Impulse Response (FIR) is implemented in the gateway in order to reduce noises from the incoming ECG signal. In order to produce more consistent, accurate, and useful information, the gateway intergate of multiple data sources is generated from different varieties of sensors; this is known as data fusion.

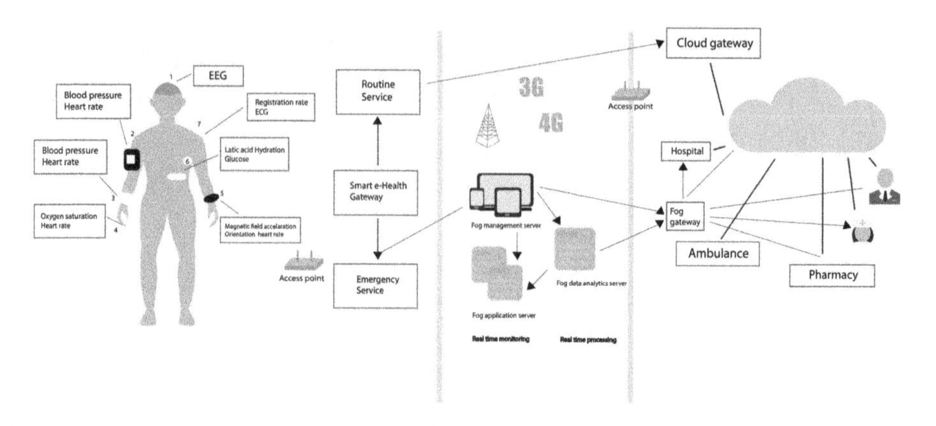

FIGURE 11.5 Three-layered architecture of EWS system.

For example, if two heart rate signals are collected from the monitored person using two different devices, but due to the presence of certain noises (movement's noises) on the received signals, the two obtained data may not be the same during the monitoring. To minimize the noise effect, first, using an anomaly detection method, outliers and meaningless data are removed, by removing the minimum threshold value of heart rate, and zero values. The reasons for having outliers in heart rate data might be that sources are mainly because of improper probe connections, or there might be a loose connection or low conductivity between the probe and the patient's skin (low moisture). Second, a weighted average can be applied to two sources values to gain a more reliable heart rate. The weight of each data set is determined by the sensor accuracy mentioned in the sensor's datasheet.

After that, data analysis is done on the data and the real-time EWS calculation is done based on rules shown in Table 11.3, which is performed at the fog layer. To calculate the EWS score, a window of collected crucial signs are selected. A sudden variation in the patient's vital signs in Figure 11.6 demonstrates the calculated score

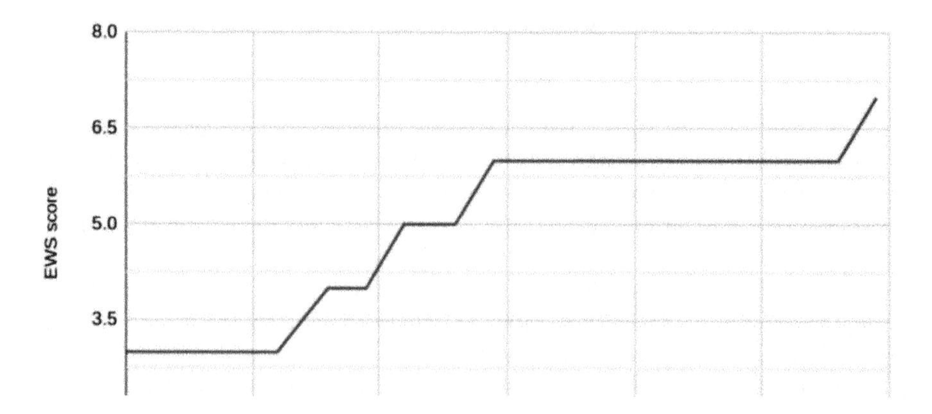

FIGURE 11.6 EWS score vs. time [42].

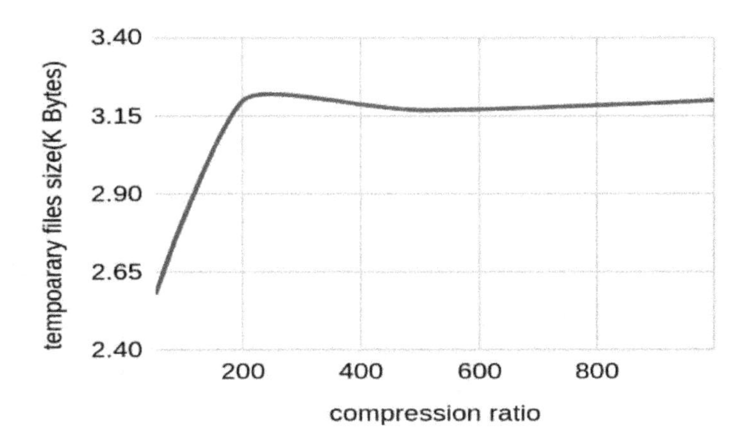

FIGURE 11.7 Compression ratio vs. temporary file size [42].

as time passes. Using a rule-based system based on Table 11.3, the score is calculated at a smart gateway. The hike in score indicates the higher chances of the patient's health deterioration for that specific window. Therefore, in this case notifications are pushed to medical experts, who are informed about the evaluative state of the patient for further attention. The sampling rate is adjusted at the gateway according to the EWS score. As the EWS score increases for patients, the sampling rate also increases, which increases the rate of sample collection to track the transitory changes more correctly. And if the EWS score is zero or in normal range, then there is a decrease in the rate of collection of samples to improve the energy efficiency of the sensors. Then data compression is done on processed data in the gateway, which is stored as a temporary file. The bigger the size of the temporary file, the higher the compression ratio would be. But there it is seen as an insignificant improvement in compression ratio for temporary files larger than 500 Kbyte, as shown in Figure 11.7 [42]. Then for security purposes, asymmetric encryption is used to keep the data safe and secure. All the compressed files on the gateway are encrypted with a public key, and only the data collector service in the cloud has the private key to decrypt the files.

At the gateway layer, there can be two different ways of addressing the data, depending on the availability of internet connection. In the first case, if the e-Health gateway has proper internet access, it does parallel real-time analysis and notification. At the same time, it tries to send the real-time data, which is non-pre-processed signals (e.g., ECG signal), to the cloud server. In this approach, the results of pre-processed output data from the first type sensor are sent along with the data from the second type and are sent to the cloud and keeps sending feedback and notifications to the sensor network in case of any alarming situation. Local storage consists of a file storage and database to keep the properties and indexes of the files. In the case of unavailability of proper internet at the gateway, it tags the unsent data to be sent in the future. A process in local storage service checks and synchronizes the stored data with the cloud server, and deletes old and replicated files from the storage, and deletes their indexes from the database, as shown in Figure 11.7.

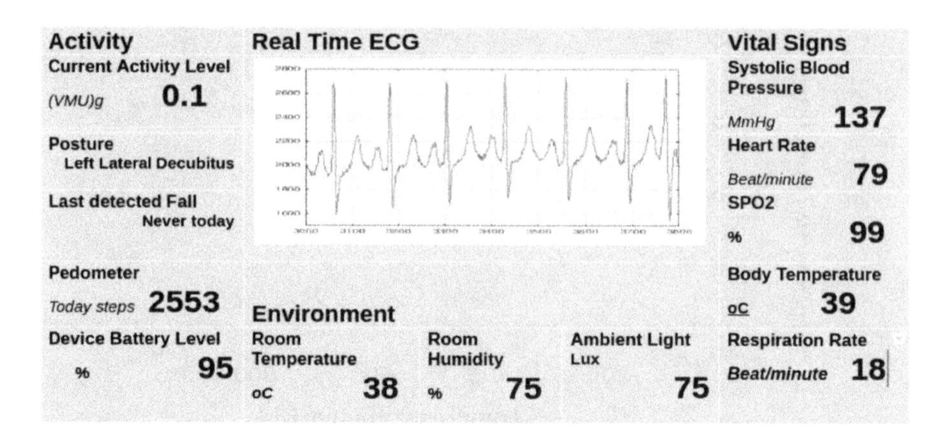

FIGURE 11.8 Live control panel web interface [48].

11.3.1.3 Backend Layer

The back-end system has two different sections:

- A cloud-based back-end infrastructure which does data storage, data analysis, decision-making, etc. The cloud server stores different types of sensor data received via the gateway and stores it in the database. On this data further analysis is done. Using the incoming data from the gateway and the patient medical history data, the server provides reports, suggestions, and possible alerts for medical experts and caregivers.
- A user interface which acts as a dashboard and performs user control and data visualization. The cloud server is also responsible for providing the administration control panel for health experts with real-time health data visualizers and a user interface for patients and in-home caregivers. The cloud server also has a user interface to provide data access for health professionals, patients, and caregivers to receive notes and notifications. In Figure 11.8 the live control panel on the web page shows various vital signs [42, 48].

11.4 CONCLUSIONS

In this chapter, the role and concept of fog computing and Smart e-Health Gateways in the context of IoT-based healthcare systems were discussed. Along with that, the challenges associated with implementation of the fog layer in an e-Healthcare system were also discussed. Also, the high-level services which can be offered by smart gateways to sensors and end-users in a geo-distributed manner at the edge of the network (e.g., local processing, storage, notification, standardization, firewall, web services, compression, etc.) were also discussed. Fog computing brings computing nearer to the user and useful in applications that are latency-sensitive and require computations to be performed near to the user. This brings a shift from clinic-centric healthcare to patient-centric healthcare. The integration of fog computing with the

IoT e-Healthcare system improved the remote patient monitoring process. Due to the introduction of the fog layer, the monitoring system was able to provide real-time analytical services, prediction and alerts in case of emergency for users and to doctors while carrying out any activity, anywhere and anytime.

The fog-based healthcare system was applied to a medical case study called Early Warning Scores, which can be used for monitoring patients with acute illnesses. In this system, patients do not necessarily need to be monitored in hospitals, and a well-structured automated solution can be utilized that will monitor patients remotely. This is possible by collecting vital signs remotely of patients' IoT sensors and IoT technologies, and hence enabling in-home patients to be monitored continuously by health experts, and they are notified when there are early signs of deterioration of health.

REFERENCES

1. Bibhash Sen, *et al.* Design of fault tolerant reversible arithmetic logic unit in QCA. *2012 International Symposium on Electronic System Design (ISED). IEEE*, December 2012. DOI:10.1109/ISED.2012.50.
2. D. Banik, Design of low cost latches based on reversible quantum dot cellular automata. *2016 Sixth International Symposium on Embedded Computing and System Design (ISED). IEEE*, 15–17 December, 2016, DOI: 10.1109/ISED.2016.7977078.
3. D. Banik, P, Bhattacharyya, and A. Ekbal, Rule-based hardware approach for machine transliteration: a first thought. *2016 Sixth International Symposium on Embedded Computing and System Design (ISED). IEEE*, vol. 12, no. 15, pp. 192–195, 2016.
4. D. Banik, Efficient methodology for testable reversible sequential circuit design. *2016 IEEE 1st International Conference on Power Electronics, Intelligent Control and Energy Systems (ICPEICES). IEEE*, July 2016. DOI: 10.1109/ICPEICES.2016.7853089.
5. D. Banik, J. Mathew, and H. Rahamant, Testable reversible latch in molecular quantum dot cellular automata framework. *2016 IEEE Annual India Conference (INDICON). IEEE*, 16–18 Dec. 2016. DOI: 10.1109/INDICON.2016.7839033.
6. D. Banik, Online testable efficient latches for molecular QCA based on reversible logic. *Design and Testing of Reversible Logic*. Springer, Singapore, 2020. pp. 185–212.
7. D. Banik, A novel approach to construct online testable reversible logic, *International Conference on Advances in Computer, Electronics and Electrical Engineering*, ISBN: 978-981-07-6260-5 2013. DOI:10.3850/978.
8. S. Mondal and D. Banik, Energy efficient fault tolerant scheme for wireless sensor networks (EEFSWSN). *IJCSET*, vol. 3, no. 3, 2013.
9. Fog computing and the internet of things: Extend the cloud to where the things are. White Paper. 2016. pp. 1–6. http://www.cisco.com/c/dam/en_us/solutions/trends/iot/docs/computing-overview.
10. S. Agarwal, S. Yadav, and A. K. Yadav, An efficient architecture and algorithm for resource provisioning in fog computing. *Int. J. Inf. Eng. Electron. Bus.*, vol. 8, pp. 48–61, 2016.
11. S. Yi, Z. Hao, Z. Qin, Q. Li, Fog computing: platform and applications. In *Proceedings of the 3rd Workshop on Hot Topics in Web Systems and Technologies, HotWeb 2015*, Washington, DC, USA, 24–25 October 2016; pp. 73–78.
12. Y. Liu, J. E. Fieldsend, and G. Min, A framework of fog computing: architecture, challenges and optimization. *IEEE Access*, vol. 4, pp. 1–10, 2017.
13. M. Mukherjee, L. Shu, and D. Wang, Survey of fog computing: Fundamental, network applications, and research challenges. *IEEE Commun. Surv. Tutor*, pp. 1826–1857, 2018.

14. M. Aazam and E. N. Huhm, Fog computing micro datacenter based dynamic resource estimation and pricing model for IoT. *Proc. Int. Conf. Adv. Inf. Netw. Appl. AINA 2015*, pp. 687–694, 2015.
15. Pirouzan Group. Available at http://pirouzansystem.com/.
16. Y. Shi, G. Ding, H. Wang, H. E. Roman, and S. Lu, The fog computing service for healthcare. In Proceedings of the 2015 2nd International Symposium on Future Information and Communication Technologies for Ubiquitous HealthCare (Ubi-HealthTech), Beijing, China, 28–30 May 2015; pp. 1–5.
17. M. Muntjir, M. Rahul, and H. A. Alhumyani, An analysis of internet of things (IoT): novel architectures, modern applications, security aspects and future scope with latest case studies. *Int. J. Eng. Res. Technol.*, vol. 6, pp. 422–447, 2017.
18. T. H. Luan, L. Gao, Z. Li, Y. Xiang, G. Wei, and L. Sun, Fog computing: focusing on mobile users at the edge. arXiv 2015, arXiv:1502.01815.
19. M. Aazam, P. P. Hung, and E. Huh, Smart gateway-based communication for cloud of things. *In Proceedings of the 2014 IEEE Ninth International Conference on Intelligent Sensors, Sensor Networks and Information Processing*, Singapore, 21–24 April 2014; pp. 1–6.
20. F. Firouzi, A. M. Rahmani, and B. Farahani, Internet-of-Things and big data for smarter healthcare: from device to architecture, applications and analytics. *Future Gener. Comp. Sys.*, vol. 78 (Part 2), pp. 583–586. DOI: 10.1016/J.FUTURE.2017.09.016.
21. W. Raghupathi and V. Raghupathi, Big data analytics in healthcare: promise and potential. *Health Sci. Sys.*, vol. 2, p. 3, 2014. DOI: 10.1186/2047-2501-2-3.
22. A. Ghazal, T. Rabl, M. Hu, F. Raab, M. Poess, A. Crolotte, and H. A. Jacobsen, Big bench: towards an industry standard benchmark for big data analytics. In *Proceedings of the 2013* ACM SIGMOD International Conference on Management of Data, pp. 1197–1208, 2013.
23. I. A. T. Hashem, I. Yaqoob, N. B. Anuar, S. Mokhtar, and A. G. Samee, The rise of "big data" on cloud computing: Review and open research issues. *Inf. Sys.*, vol. 47, pp. 98–115, 2015.
24. A. Bourouis, A. Zerdazi, M. Feham, and A. Bouchachia, M-Health: skin disease analysis system using smartphone's camera. *Procedia Comput. Sci.*, vol. 19, pp. 1116–1120, 2013.
25. N. Choi, D. Kim, S. Lee, and Y. Yi, Fog operating system for user-oriented IoT services: challenges and research directions. *IEEE Commun. Mag.*, vol. 55, pp. 2–9, 2017.
26. M. Gordon, R. Henderson, J. H. Holmes, M. K. Wolters, and I. M. Bennett, Participatory Design of ehealth solutions for women from vulnerable populations with perinatal depression. *Am. Med. Inform. Assoc.,* vol. 23, no. 1, pp. 105–109, 2016 January. DOI: 10.1093/jamia/ocv109.
27. G. Suciu, V. Suciu, A. Martian, R. Craciunescu, A. Vulpe, I. Marcu, S. Halunga, and O. Fratu, Big Data. Internet of things and cloud convergence – an architecture for secure e-health applications. *J. Med. Sys.*, vol. 39, no. 141, 13 Nov. 2015.
28. R. Roman, J. Zhou, and J. Lopez, On the features and challenges of security and privacy in distributed internet of things. *Comp. Netw.*, vol. 57, pp. 2266–2279, 2013.
29. Symantec. Available at https://www.symantec.com/solutions/internet-of-things.
30. WindRiver. Available at http://www.windriver.com/iot/.
31. Q. Jing, A. V. Vasilakos, J. Wan, J. Lu, and D. Qiu, Security of the internet of things: perspectives and challenges. *Wirel. Netw.*, vol. 20, no. 8, pp. 2481–2501, 2014.
32. J. Ni, K. Zhang, X. Lin, and X. Shen, securing fog computing for internet of things applications: challenges and solutions. *IEEE Commun. Surv. Tutor,* vol. 20, pp. 601–628, 2017.
33. N. K. Giang, M. Blackstock, R. Lea, and V. C. M. Leung, Developing IoT applications in the fog: a distributed dataflow approach. In *Proceedings of the 5th International Conference Internet Things,* IoT 2015, Seoul, Korea, 26–28 October 2015; pp. 155–162.

34. M. Ketel, Fog-cloud services for IoT. In *Proceedings of the SouthEast Conference*, Kennesaw, GA, USA, 13–15 April 2017; pp. 262–264.
35. Fog computing and the internet of things: extend the cloud to where the things are. White Paper. 2016. http://www.cisco.com/c/dam/en_us/solutions/trends/iot/docs/computing-overview.pdf.
36. H. F. Atlam, A. Alenezi, A. Alharthi, R. Walters, and G. Wills, Integration of cloud computing with the internet of things: challenges and open issues. In *Proceedings of the 2017 IEEE International Conference on Internet of Things (iThings) and IEEE Green Computing and Communications (GreenCom) and IEEE Cyber, Physical and Social Computing (CPSCom) and IEEE Smart Data (SmartData)*, Exeter, UK, 21–23 June 2017; pp. 670–675.
37. O. Anya and H. Tawfik, Designing for practice-based context-awareness in ubiquitous e-health environments. *Comput. Elect. Eng.*, vol. 61, pp. 312–326, 2017.
38. J. Granados, A. M. Rahmani, P. Nikander, P. Liljeberg, and H. Tenhunen, Towards energy-efficient Health Care: An internet-of-things architecture using intelligent gateways. In *Proceedings of International Conference on Wireless Mobile Communication and Healthcare*, 2014, pp. 279–282.
39. B. Farahani, F. Firouzi, V. Chang, M. Badaroglu, N. Constant, and K. Mankodiya, Towards fog-driven IoT eHealth: Promises and challenges of IoT in medicine and healthcare. *Future Gener. Comp. Sys*, vol. 78, no. 2, pp. 659–676. DOI: 10.1016/j.future.2017.04.036.
40. H. F. Durrant-Whyte, Sensor models and multisensor integration. *Int. J. Robot. Res.*, vol. 7, pp. 97–113, 1988.
41. Health Level Seven Int'l. Introduction to HL7 Standards, 2012. Available at www.hl7.org/ implement/standards.
42. A. M. Rahmani, T. N. Gia, B. Negash, A. Anzanpour, I. Azimi, M. Jiang, and P. Liljeberg, Exploiting smart e-Health gateways at the edge of healthcare internet-of-things: a fog computing approach. *Future Gener. Comp. Sys*, pp. 1–46, February 2017. DOI: 10.1016/j.future.2017.02.014.
43. netfilter/iptables-nftables project, http://netfilter.org/projects/nftables/.
44. S. R. Moosavi, T. N. Gia, A. Rahmani, E. Nigussie, S. Virtanen, H. Tenhunen, and J. Isoaho, SEA: A secure and efficient authentication and authorization architecture for IoT-based healthcare using smart gateways. In *Proceeding of 6th International Conference on Ambient Systems, Networks and Technologies*, 2015, pp. 452–459.
45. S. R. Moosavi, T. N. Gia, E. Nigussie, A. Rahmani, S. Virtanen, H. Tenhunen, and J. Isoaho, Session resumption-based end-to-end security for healthcare internet of-things. In *Proceedings of IEEE International Conference on Computer and Information Technology*, 2015, pp. 581–588.
46. C. Otto, A. Milenković, C. Sanders, and E. Jovanov, System architecture of a wireless body area sensor network for ubiquitous health monitoring, *J. Mob. Multimedia*, vol. 1, no. 4, pp. 307–326, 2006.
47a. T. Nguyen Gia, N. K. Thanigaivelan, A. M. Rahmani, T. Westerlund, P. Liljeberg, and H. Tenhunen, Customizing 6LoWPAN networks towards Internet-of-Things based ubiquitous healthcare systems, *IEEE*, vol. 10, no. 27, pp. 1–6, 2014.
47b. J. Xiaoying, H. Debiao, N. Kumar, and K-K. R. Choo, Authenticated key agreement scheme for fog-driven IoT healthcare system Wireless Networks volume. *Comp. Sci.*, 2019. DOI: 10.1007/S11276-018-1759-3.
48. A. Anzanpour, A.-M. Rahmani, P. Liljeberg, and H. Tenhunen, Internet of things enabled in-home health monitoring system using early warning score, *ACM International Conference on Wireless Mobile Communication and Healthcare*. January 2015, DOI: 10.4108/eai.14-10-2015.2261616.

49. A. Ukil, J. Sen, and S. Koilakonda, Embedded security for internet of things. *2011 2nd National Conference on Emerging Trends and Applications in Computer Science*, 4–5 March 2011. *IEEE*. DOI: 10.1109/NCETACS.2011.5751382/
50. O. C. Omeni *et al.* Energy efficient medium access protocol for wireless medical body area sensor networks. In *Medical Devices and Biosensors*, pp. 29–32, Aug. 2007.
51. U. Kyriacos *et al.* Monitoring vital signs using early warning scoring systems: a review of the literature. *J. Nurs. Manage.*, vol. 19, no. 3, pp. 311–330, 2011.
52. M. Odell, C. Victor, and D. Oliver, Nurses' role in detecting deterioration in ward patients: systematic literature review, *J. Adv. Nurs.*, vol. 65, no. 10, 2009.
53. T. O'Kane *et al.* Mews to e-mews: From a paper-based to an electronic clinical decision support system. In European Conf. on Information Management, pp. 301–305, January 2010.
54. J. McGaughey *et al.* Outreach and early warning systems (EWS) for the prevention of intensive care admission and death of critically ill adult patients on general hospital wards. *Cochrane Database Syst. Rev.*, vol. 3, 2007.
55. C. Franklin *et al.* Developing strategies to prevent in hospital cardiac arrest: analyzing responses of physicians and nurses in the hours before the event. *Critical Care Medicine*, vol. 22, no. 2, pp. 244–247, February 1994.
56. R. M. Schein *et al.* Clinical antecedents to in-hospital cardiopulmonary arrest. *Chest*, vol. 98, no. 6, pp. 1388–1392, December 1990.
57. D. Georgaka *et al.* Early warning systems. *Hospital Chronicles*, vol. 7 (1 Sup), pp. 37–43, 2012.

12 IIoT: A Survey and Review of Theoretical Concepts

Souptik Ghosh, Mahendra Kumar Gourisaria, Siddharth Swarup Routaray, Manjusha Pandey
School of Computer Engineering, KIIT Deemed to be University, Bhubaneswar, Odisha, India

CONTENTS

12.1 Introduction 223
12.1.1 The Industrial Internet of Things (IIoT): An Analysis Framework 224
12.1.2 Large-Scale IIoT Deployments 224
12.1.3 Development of IIoT in Industrial Product Service Systems 225
12.1.4 IIoT Integration in Learning Factory 227
12.1.5 IIoT and the Digital Society 228
12.2 Literature Survey 229
12.3 Discussion and Review 231
12.3.1 The Industrial Internet of Things (IIoT): An Analysis Framework (2018) 231
12.3.2 Large-Scale IIoT Deployments (2019) 231
12.3.3 Development of IIoT in Industrial Product Service Systems (2018) 231
12.3.4 IIoT integration in Learning Factory (2019) 231
12.3.5 IIoT and the Digital Society (2016) 231
12.4 Case Studies and Future Directions for IIoT 232
12.4.1 A Waste-to-Energy Plant 232
12.4.2 IIoT in Logistics 232
12.4.3 IIoT in Underground Coal Mines 233
12.4.4 Security Automation with IIoT 233
Acknowledgement 234
References 234

12.1 INTRODUCTION

IIoT will help to create new business references by improving productivity, by using analytic studies for creation, and transforming the work power. While connectivity and data aggregation are of utmost importance for Industrial Internet of Things (IIoT),

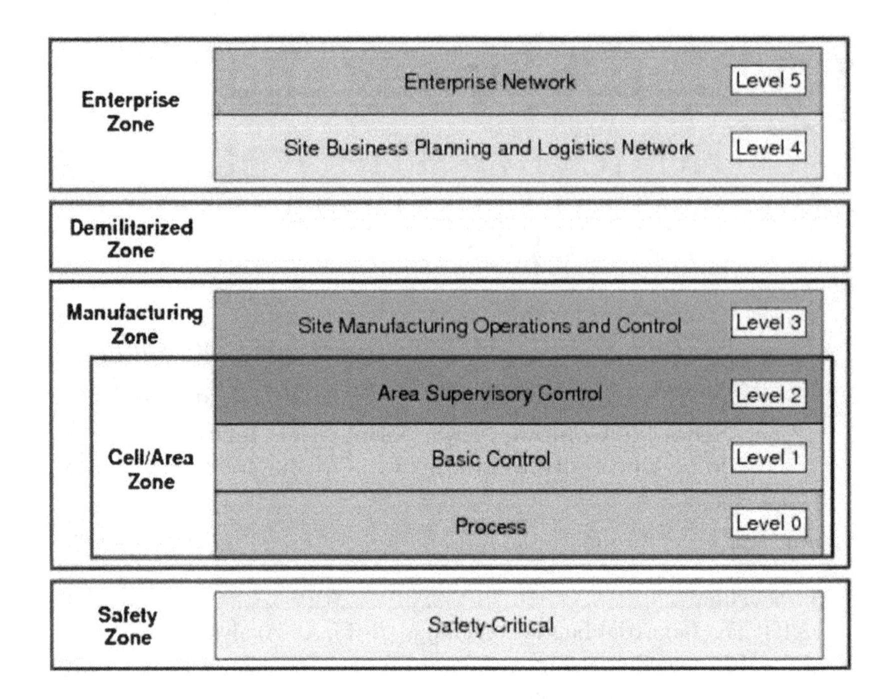

FIGURE 12.1 IoT solutions structure depicted in four stages.

predictive analysis and maintenance is the true application. Some of the important work done on IIoT has been analyzed in the presented review paper from credible sources, such as papers of different authors published on journals, such as Elsevier, De Gruyter, author manuscripts in PMC-NCBI, IJEE, etc. This paper was done with the aim to review the previous work done in this field and to help the progress of the future work to be done in this field.

12.1.1 The Industrial Internet of Things (IIoT): An Analysis Framework

The idea of Industrial Automation and Control Systems (IACS) is well known [1]. These processes, usually known as Operational Technology (OT), are used in different industrial processes such as manufacture, transport, and utilities. Also, sometimes these are known as cyber-physical systems (CPSs) or jointly as Industrial Internet of Things (IIoT) (see Figures 12.1 and 12.2).

12.1.2 Large-Scale IIoT Deployments

Mammoth manufacturing systems require usage of actuators and sensors to the existing machinery, as well as installing new ones with integrated updated sensors [2]. Owing to the complex properties of the network, the sensor data do not flow smoothly and continually as wanted. The sensor data may be not found, due to network problems such as congestion, time lag, sensory mechanism failure, and sensor data change. The data of that sensor on the failure period needs to be found and

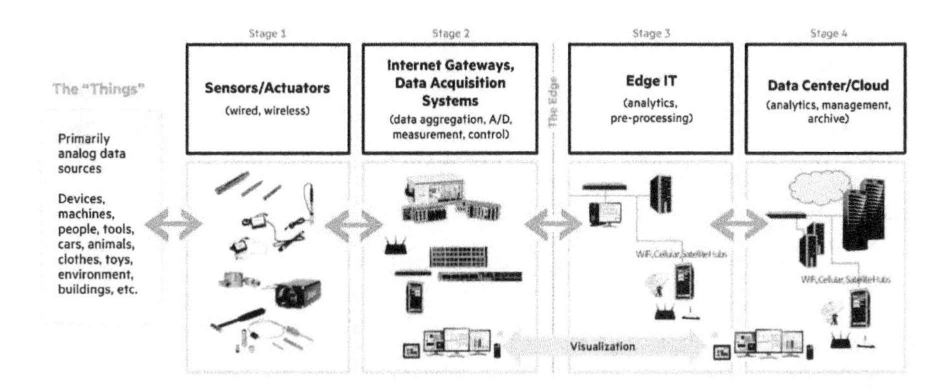

FIGURE 12.2 Purdue reference model.

hence is the concept of large-scale deployments of such interconnected sensors (see Figures 12.3 and 12.4).

12.1.3 Development of IIoT in Industrial Product Service Systems

Manufacturing industries have changed radically during the last few decades, and the transfer of percentage profit from goods to systems has morphed conventional

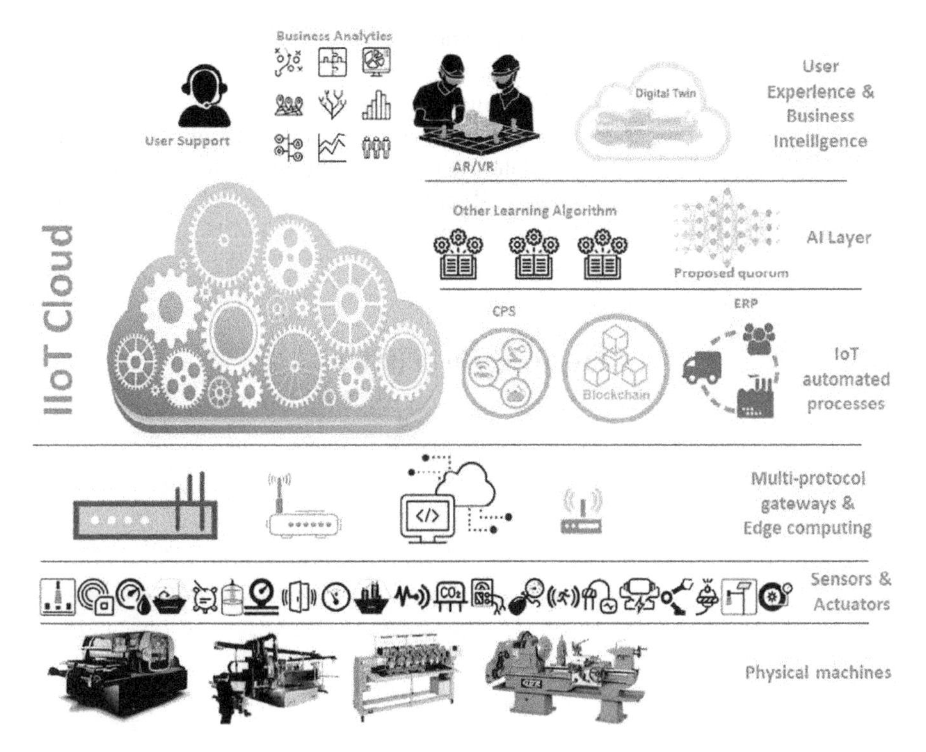

FIGURE 12.3 Structure of the wanted MLQD system.

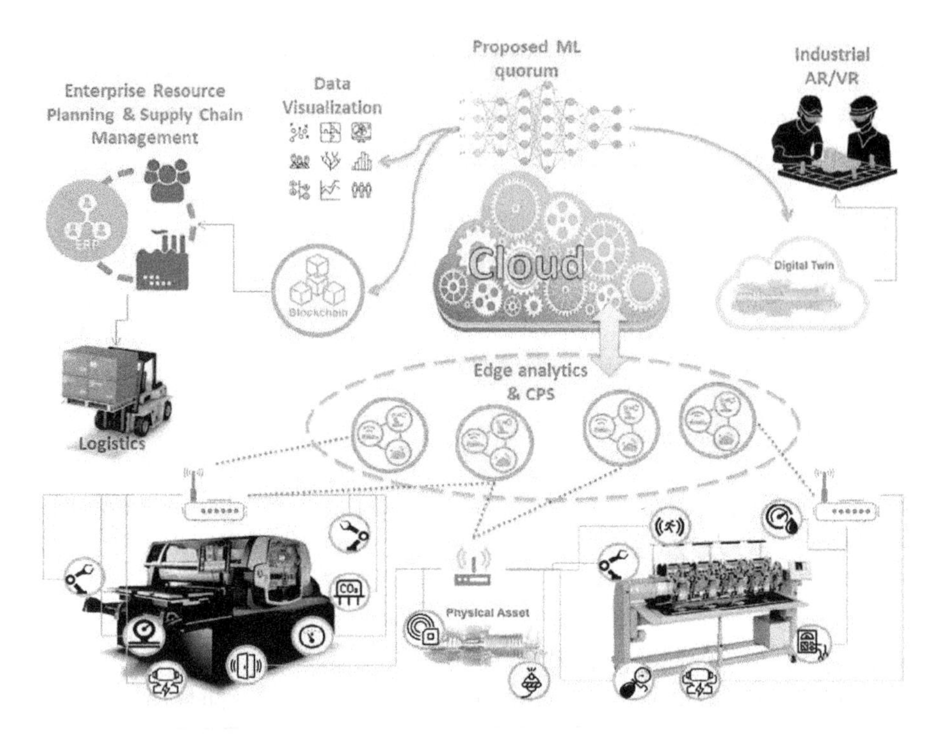

FIGURE 12.4 Work process of the MLQD Frameworks.

equipment of production from supplier industry to distributors of Industrial Product-Service Systems (IPSS) [3]. IPSS is a new business model of Industrial Product-Service Systems (IPSS). IPSS is a latest commerce venture for the consignment of industrial goods and systems where (IoT) machines are integrated into the product for supervision, regulation, and management (see Figure 12.5 to 12.7).

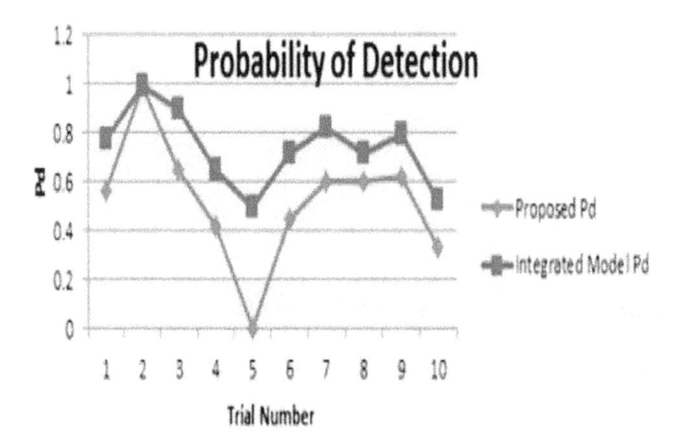

FIGURE 12.5 Prospect of sensor failure perception.

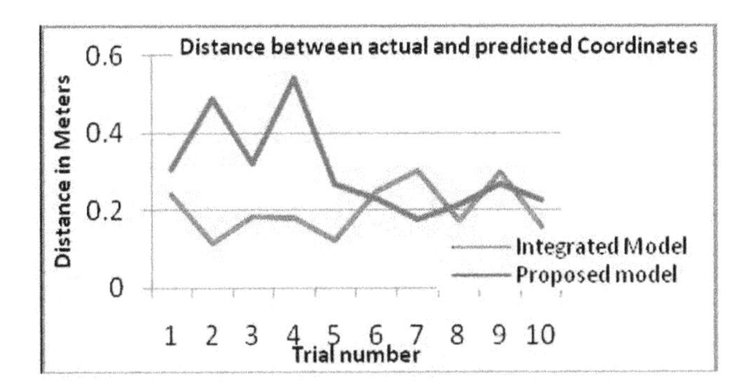

FIGURE 12.6 Probable traced coordinates vs. real coordinates.

12.1.4 IIoT Integration in Learning Factory

The proposed foundations of the future factory are ones in which digital technology may be revolutionary for manufacturers [4]. These mechanizations could similarly affect other sectors, such as health, transport, and environmental concerns, equally. A few of the new advancements in AI, automation, IOT, and the development of 5G networking systems are shifting the world for Industry 4.0. LF (Learning Factories) has an immense role in this transformation and is a required, hands-on experience gathering opportunity for students. IIoT and CPS have been embedded in the LF with their implementation in product concept to finished products. Continuous improvement of the LF is thus mandatory for sustained growth and development (see Figure 12.8).

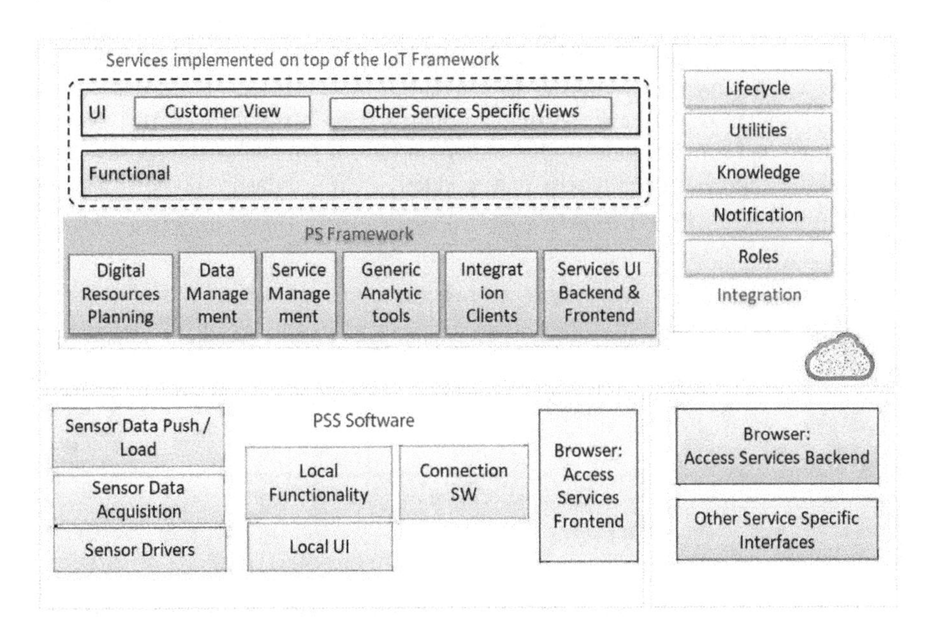

FIGURE 12.7 Services framework reference architecture.

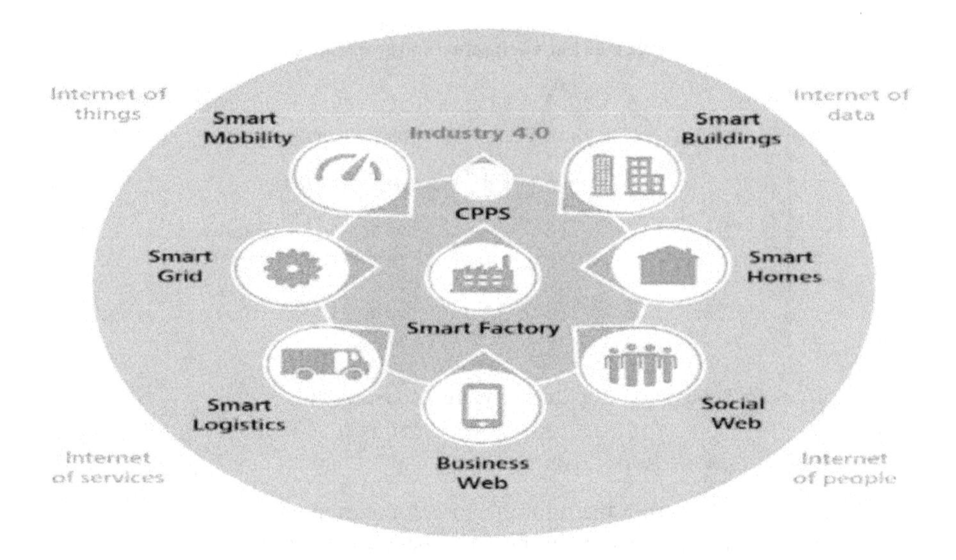

FIGURE 12.8 RFID lead flow operation for manufacturing solenoid valves.

12.1.5 IIoT and the Digital Society

Due to the rapid development of communication and IT, we live today in a networked digital platform, characterized with increased spontaneity and immediate gain of data [5]. Simultaneously, we must accommodate with a vast number of global concerns, such as environment protection, improvement of health, etc. Companies must adopt increasing consumer-friendly approaches, aiming to the aggregated achievement of commercial growth and sociological progress for continued growth. The environment has pressured industry to respond to these challenges and risks while taking advantage of the newly available opportunities in the form of IIoT and its applications and benefits. Industry has undergone a series of "industrial revolutions," and with the increasing complexity and productivity, it is industry 4.0 now that's of utmost importance (see Figure 12.9).

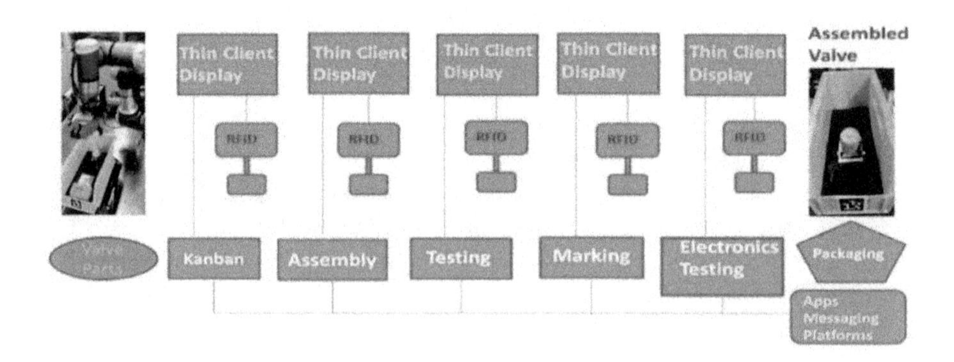

FIGURE 12.9 Industry 4.0 environment.

12.2 LITERATURE SURVEY

Sl. No.	Year/Author	Title	Proposal	Advantage	Disadvantage
1	2018: Hugh Boyes, Bil Hallaq, Joe Cunningham, Tim Watson.	The Industrial Internet of Things (IIoT): An analysis framework [1].	1. Assessing the IIoT ecosystem and risk prospect. 2. Increased research in OT and ICT confluence. 3. Giving answer to Brownfield problems. 4. Research on safety and security of IIoT.	1. Better identification of the vulnerabilities and potential security concerns. 2. Better informed assessment accounting for the short coming of IIoT machines.	1. No solutions to Brownfield issues at present; further research is required. 2. Currently there is a lack of persistent undertaking of the integrated assessment of security and safety risks inherent in usage of IIoT solutions.
2	2019: Ariharan V, Subha P. Eswaran, Srinivasarao Vempati, Naveed Anjum.	Large scale IIoT deployments [2].	A special system of network called the MLQD, which gives predictions from AI algorithms to take final required actions.	1. High-level intelligence provides data predictions even during loss of sensor values due to byzantine failure, thus operational flow is uninterrupted.	Cumbersome voting analysis and rendering of incorporated algorithms lowers more extensively than the theoretical algorithm.
3	2018: Kosmas Alexopoulos, Spyros Koukas, Nikoletta Boli, Dimitris Mourtzis.	Development of IIoT in industrial product service systems [3].	Developing an IIoT model that backs the sense, undertaking, and operational states of services in IPSS, which is a new commerce prototype for reliable delivery of products and services to industries.	1. IPSS with analysis and accumulation of feed from laser equipment. 2. NoSQL databases support high-scale and highly available feed without compromising the rendering requisites.	1. Here the proposed framework consists mostly of edge sensors in which data aggregation takes place, hence causing data overflow in the processers. 2. A few of the latest high-tech production machines are able to store data concerning the utilization of the machinery, identifying the usage of the machine in use, and hence can cause failures as alarm events, internal failures, and abnormal outputs.

4	2019: Ishwar Singh, Dan Centea, Mo Elbestawi.	IIoT integration in Learning Factory [4].	The aim for the application of LF WBHMI is so that an aggregated IoT and IIoT structure is received using HTML5 and WS protocol.	Conversion to TCP/IP or MODBUS is easy, hence another machine that supports transmission through the TCP/IP process can as well be embedded with these systems with PLCs, DCSs, etc.	The heart of CPS is the SEPT-LF having the penta axillary CNC device but not supporting the MT Connect transmission level, thus ethernet connection is the only option connecting the machine to a TCP server.
5	2016: Luis Miguel Fonseca	IIoT and the digital society [5].	Overview of the several industrial revolutions and IIoT with its underlined dimensions.	1. Shortened "time to market" to develop, produce, and market new products and services, requiring higher and faster innovation capability. 2. Increased customization to satisfy individual consumer demands. 3. Higher flexibility with faster and more versatile production processes able to produce smaller lot quantities with high precision and cost efficiency. 4. Increased resource efficiency by using more efficient and closed loops.	1. Creates higher worth connections and changes the traditional person and manufacture organization systems, impacting society as overall. Thus, pointing for the requirement of an escalated system induction, which means that production organizations will not be able to focus only on their products any more. 2. Employees with skill level-based jobs risk replacement, as most of the manual labor-intensive jobs and manual monitoring will be replaced by AI machines, capable of self-control and coordination, along with precision. Monitoring will also be taken over by machines and steps will be automated.

12.3 DISCUSSION AND REVIEW

12.3.1 The Industrial Internet of Things (IIoT): An Analysis Framework (2018)

The identification of the problems in the different fields is very accurate and the solutions provided are quite apt [1]. The idea of segregation of each component under different fields, as well as each field having multiple components, is unique. However, the assumptions made are quite bold and hence can be in theory only; thus, practical applications are still far off.

12.3.2 Large-Scale IIoT Deployments (2019)

High-intelligence predictions, along with uninterrupted operational flow, are the need of the hour, but reliability and sustained integrated performance are also required [2, 6, 7]. The missing gaps in data can be overcome by using AI, but actual data are still far more reliable, and if unprompted errors occur, then it will go on uncounted for, which can cause problems later.

12.3.3 Development of IIoT in Industrial Product Service Systems (2018)

This work provides an IIoT structure for accomplishing a system of output in IPSS [3, 8]. The model gives varied function-aligned output for separate user parts and each individual requirement. However, to deploy this framework utilization and testing over an adequate period of time is required, which may be too lengthy and hence not appropriate.

12.3.4 IIoT integration in Learning Factory (2019)

This SEPT LF is one service that gives a new visual of prospected manufacturing companies, but again they also give students of engineering with a straight perspective to better themselves and their abilities [4, 9, 10]. They then acquire fundamental theories by implementing IIoT and IoT learning prototypes. Later, they will have the chance to be included in various projects like innovating useful apps using IIoT and IoT devices, improving and designing existing processes. The process is very useful and simple, but not very referenceable to actual practical applications, and does not show any out-of-circumstance errors or troubles that would be helpful for the future.

12.3.5 IIoT and the Digital Society (2016)

IIoT means the digitalization of production via interconnected systems of persons and machines working and interacting together [5, 11]. The aim is for companies, by effectively implementing IIoT, to considerably flourish their position in the competitive market, increasing valued innovation and lessening threats, with the implementation of better and swifter manufacturing processes and newer technologies. Yet if the already working class is not properly trained to handle the machines, they themselves may face replacement and hence unemployment and its effects may see a rise.

12.4 CASE STUDIES AND FUTURE DIRECTIONS FOR IIoT

12.4.1 A Waste-to-Energy Plant

Usage of data analytics for prediction in objective organizations, using delicate sensors, are precious instruments for giving insights into the process of operation and workings, especially for a case where the straight assessment of fundamental process fluctuations are very problematic, almost unfeasible, or even inconsistent [12, 13]. One such place is the waste-to-energy plant. Here the Soft sensors are generated by embedding mathematical operation prototypes. These give an economical replacement to practical measuring machines and actual-time prediction of objective fluctuations, which may be simply used along with or embedded with the measuring devices. This type of sensor is able to handle industrial Big Data, resulting in escalated variable projection. All the thousands of sensors, their supervising modules, the coordinating connections, and the cloud, connecting all of these, together makes up the whole plant. However, the optimization of such plants is mandatory and has not yet been fully achieved. As a result, sensors, which are very high in numbers, will flood the actuators with such a huge amount of data that it will be impossible for it to relay it forward, and thus, a potential breakdown of the whole system may be an imminent threat. Moreover, a high number of sensors also poses for cases of malfunction and hence replacement. But detection and positioning of them will be a taxing job, and hence may remain unattended, and again the hierarchy may face collapse. A collapse of even a part of this intricate hierarchy will cause failure of the machines and hence failure to attain the actual goal.

12.4.2 IIoT in Logistics

The requirement for highly singulated services and products is increasing continuously [14–17]. Hence, outbound and inbound goods have to adjust to this reforming circumstance. Due to its aggravating intricacy, it can't be operated with standard planning or controlling procedures. Implementation of IIoT in logistics, which in CPS-based services is known as smart services and technology, makes up the smart logistics. These smart logistics are systems, which are able to improve the adaptability for the adjustments to the changes in the markets, and will, in turn, result in the companies being closer to the needs of the customers. These will make it possible to uplift the levels of services to the customer, the development of the manufactured product, and will make the prices of storing and manufacturing low. Smart products and services are ones which can undertake tasks that are usually executed by humans, so the human work power can be focused on the jobs that need better perception than automated systems, or the intellect, which simple smart products or smart services can deliver. This will enhance the overall production, adaptability, and swiftness to the transformations, which may take place in the supply networks. Correct inclination and incorporation among the key components of the supply network, and increasing degrees of apprehension and clarity, will make sure a sufficient forecast of assets and hence optimize production. However, the large number of processes that are maintained by human skill and experience is very difficult to be received by machines and automation. At just the present stage of implementation, it is almost impossible to achieve the amount of perfection already at work by the human resources and will cause loss to products and service quality as well.

12.4.3 IIoT in Underground Coal Mines

IIoT is an idea which fuses control systems and sensor networks and has been put to use in various professions to better the manufacturing abilities and safety [18–20]. NIOSH researchers from the United States have been evaluating applications of IIoT for IIoT implementation in the mining industry. IIoT is expected to possess the ability to reform the mining systems for years to come. Along with the progresses in sensor techniques, additional sensors are being employed in mines to measure operational factors and environmental statistics of vitality. The mining organizations have been employing PLC and SCADA systems for decades. But they were not interconnected and could not pass information among themselves. IIoT structures are basically an open, interconnected (IP) network architecture. Transformation to a better open network structure and feed transfer of IIoT gives opportunities to an industry such as mining, which will make it able to advance the future generations of automation and smarter mining. Faster relay of information, quick preventive measures, and safer mining are no doubt the greatest possibilities of IIoT implementation in mining. Yet problems are not far off. To properly connect a place as extensive and vast as a coal mine, it is given that a huge number of sensors are to be used and a similar huge number of connectors. Such devices are built to work under sophisticated conditions and not in the abrasive conditions of a coal mine. This will cause regular failures, and changes will not be possible. Moreover, the huge amount of data relayed by the sensors to be processed may cause a data overload and hence cause the system to malfunction.

12.4.4 Security Automation with IIoT

Appliance automation is one of the most significant developing advances that can change the way people live in the world [21–25]. Automation systems available in the market use different wireless communications, enabling the appliances to be turned on or off using a remote control instead of going to the switchboard. Easier interface and ease of control, along with the possibility of revolutionizing beautification, has caused IIoT to be implemented in security and automation, but the situational analysis and immediate decision-taking abilities shown by humans are very difficult to be received from machines, and initiative-taking situations are to the most affected parts. Some common problems that can arise are shown in Table 12.1.

TABLE 12.1
Security Requirements and Challenges for IIoT

N.O.	Security Requirements	Security Challenges
1	Confidentiality: The process of keeping something private.	Computational limitations
2	Authentication: The process of proving something authentic.	Memory limitations
3	Self-healing: The process of recovery of the body itself.	Mobility
4	Fault tolerance: The process of continuing proper operations even during failure.	Scalability
5	Resiliency: The process of ability to recover quickly from difficulty.	Communications gaps
6	Availability: The process of being able to be used as and when required.	A multi-protocol network

ACKNOWLEDGEMENT

I am grateful to all my co-authors and mentors who have guided and helped me throughout this project to bring it into perfection. I also thank KIIT (deemed to be university) School of Computer Sciences for giving me this opportunity to do this article.

REFERENCES

1. Boyes H., Hallaq B., Cunningham J., Watson T. (2018). The industrial internet of things (IIoT): an analysis framework. Computers in Industry, 101, 1–12.
2. Ariharan V., Subha P.E., Srinivasarao V., Naveed A. (2019). Large scale IIoT deployments. Procedia Computer Science, 151, 959–964.
3. Kosmas A., Spyros K., Nikoletta B., Dimitris M. (2018). Development of IIOT in industrial product service systems. Procedia CIRP, 72, 880–885.
4. Singh I., Centea D., Elbestawi M. (2019). IIOT integration in learning factory. Procedia Manufacturing, 31, 116–122.
5. Luis M.F. (2016). IIOT and the digital society. Industry 4.0 and the digital society: concepts, dimensions and envisioned benefits. PICBE-2018-0034, pp. 386–397.
6. Hamel G., Välikangas L. (2003). The quest for resilience. Harvard Business Review, 81(9), 52–63.
7. Hermann, M., Pentek, T., Otto, B. (2016). Design principles for Industrie 4.0 scenarios, in: 2016 49th Hawaii International Conference on System Sciences (HICSS), IEEE, pp. 3928–3937.
8. Gobierno de España (2016). La transformación digital de la industria española, Industria conectada 4.0, accessed 2017.10.25.
9. Drath R., Horch H. (2014). Industrie 4.0: hit or hype? [industry forum]. IEEE Industrial Electronic Magazine, 8(2), 56–58.
10. Fonseca L., Ramos A., Rosa A., Braga A.C., Sampaio P. (2016). Stakeholders satisfaction and sustainable success. International Journal of Industrial and Systems Engineering, 2 (2), 144–157.

11. Bagheri B., Yang S., Kao H.A., Lee J. (2015). Cyber-physical systems architecture for self-aware machines in industry 4.0 environment. IFAC-PapersOnLine, 48, 1622–1627.
12. James C.K., Sirkka L Jämsä J., Robert S., Christian B. (2019). Industry 4.0 based process data analytics platform: a waste-to-energy plant case study. Electrical Power and Energy Systems, 115, 105508.
13. Davies R. (2015). Smart Factory Market. Briefing September 2015 (R. EPRS, European Parliamentary Research Service. Members' Research Service PE 568.337).
14. Barretoa L., Amarala A., Pereira T. (2017). Industry 4.0 implications in logistics: an overview. Procedia Manufacturing, 13, 1245–1252.
15. Chesbrough H. (2006). Open Business Models: How to Thrive in the Innovation Landscape. Boston, MA: Harvard Business School Press.
16. Chesbrough H. (2007). Business model innovation: It's not just about technology anymore. Strategy and Leadership, 35(6), 12–17.
17. Cippola C.M. (1965). Guns, Sails, and Empires: Technological Innovation and the Early Phases of European Expansion, 1400-1700. Sunflower University Press. ISBN-10: 089745071X.
18. Zhou C., Damiano N., Whisner B. Reyes M. (2017). IIOT applications in underground coal mines. Author manuscript in PMC 2018 January 16.
19. European Commission. (2015). Digital transformation of European industry and enterprises, Strategic Policy Forum on Digital Entrepreneurship.
20. Lorenz M., Rüsmann M., Waldner M., Engel P., Harnisch M., Gerbert P., Justus J. Industry 4.0: the future of productivity and growth in manufacturing industries. Boston Consulting Group, 2015.
21. Devi G (2019). Security aware intelligent automation using internet of things (IOT). IJEE, 11, 1, 62–67.
22. Amit R., Zott C. (2012). Creating value through business model innovation. MIT Sloan Management Review, 53(3), 41–49.
23. Arnold, Christian. (2017). The industrial internet of things from a management perspective: A systematic review of current literature. Journal of Emerging Trends in Marketing and Management, 1 (1), 8–21.
24. Bucherer E., Eisert U., Gassmann G. (2012). Towards systematic business model innovation: Lessons from product innovation management. Creativity and Innovation Management, 21(2), 183–198.
25. Casadesus-Masanell R., Zhu F. (2013). Business model innovation and competitive imitation. Strategic Management Journal, 34(4), 464–482.

Index

A

Actuator, 27, 101, 204
Agriculture, 6, 8, 17, 72, 82, 94, 95
AHP, 171–174, 176, 179
Alzheimer's, 20
Ambient, 5, 26, 72, 85
AOA, 173
Application layer, 3–6, 24, 26, 73, 76, 100, 143, 158, 204
Authentication, 3, 21, 36, 146, 202, 211, 234
Authorization, 3, 36, 122, 131, 205, 210
Automation, 54, 55, 82, 110, 117, 137, 224

B

BASN, 19
Big Data, 6, 7, 11, 22, 37, 62, 202, 206, 211, 214
BioMedical sensors, 22
BLE, 32, 201, 211
Blood pressure sensor, 30, 146
Bluetooth, 32, 51, 61, 76, 94, 140, 201
Body temperature sensor, 31, 146
BSN, 22
Business layer, 5, 6, 24, 34, 173
Business models, 6, 49, 173

C

Carousel attack, 128, 129, 228
Chemical sensor, 26, 228
Cloud computing, 6, 21, 29, 34, 37, 38, 93, 97, 154, 196
Cloud server, 3, 12, 52, 56, 148, 196, 205, 212, 217, 218
Complexity, 25, 48, 51, 132, 198, 209, 228
Congestion, 10, 66, 108, 157, 161, 163, 165, 224
Content integrity, 3
CPS, 227, 228, 230

D

Data analysis, 22, 24, 37, 49, 61, 63, 84, 200, 203, 216, 218
Data cleaning, 37
Data link layer, 3, 128
Data plane, 125
Dipole, 4, 36
Data storage, 4, 55, 59, 60, 122, 123, 218
Denial of service, 39, 120, 126, 130, 211
Distributed computing, 1, 36, 93, 197, 198, 206
Dynamic nodes, 39

E

ECG, 22, 30, 143, 205, 209, 214, 217
Edge layer, 58, 61, 204
EEG, 30, 203, 205
Electronic sensors, 2
EMG, 30, 203
Encryption, 7, 122, 123, 205, 211, 217
Energy efficient, 8, 21, 134, 155, 159
Energy harvesting techniques 12, 140, 159
Event handler 60, 204

F

Fault tolerance 48, 78, 132, 211
Fog computing/layer 195–198, 207, 208, 211
Force sensor 26
FSONA 172, 173, 175, 179, 184

G

Galvanic skin response, 30
Game theory, 118
Gateway, 26, 28, 52, 59, 61, 68, 96, 203, 205, 208, 213, 215, 217
Geomagnetic sensors, 159
Gesture recognition, 30
Glucometer, 30
Glucose sensor, 31
GPS, 10, 17, 74
Green buildings, 168
Green technology, 8
Greenhouse gas, 155, 159, 164
Grid computing, 1
GSM, 35, 205
GUI, 183, 185
Gyroscopes, 10

H

Habitat monitoring, 4, 137
Hacker, 124
Health monitoring, 4, 9, 10, 20, 139, 195, 204, 211, 212, 213
Health-IoT system, 39, 200, 203–206
Heart rate, 22, 30, 31, 200, 201, 214, 216

Heterogeneous, 7, 9, 46, 117, 131, 137, 139, 141, 142, 154, 173, 197, 205
Home automation, 4, 8, 117
Home automation system, 117
Household devices, 117
Hybrid communication, 92
Hypergraph, 100, 106, 108, 110
Hypoxia, 31

I

ICMP, 128
IEC, 64
Information and communication theory, 1, 72, 204
Infrared, 6, 10, 26
Infrastructure as a service, 36, 143
Integrity, 3, 7, 8, 65
Intelligent transportation systems, 10, 117
Interface, 3, 28, 36, 46, 59, 61, 77, 80, 122
Internet Engineering Task Force, 122
Intruder, 121, 128
IP addresses, 118, 120, 134
ISM, 4
ITU, 1, 64, 94

J

Jamming, 121, 127, 201, 210
JSON, 47, 62, 147

K

Kernel, 205
KiORH application, 137, 145, 146, 149
Kiosk, 137, 138, 144–146, 149

L

LAN, 26, 134
Latency, 49, 52, 196, 199, 202, 205
Listener, 144
Localization, 74, 118, 157, 172, 173, 176, 193
LoRa, 34
Low Power Wide Area Networks, 33
LSS, 100, 105, 106
LTE, 12, 21, 35

M

Machine learning, 5, 11, 12, 37, 40, 58, 60, 63, 93, 134, 203, 207, 214
Malicious, 65, 95, 119, 120, 122, 125, 127, 128, 130, 133, 134, 211
Man-in-the-middle attack, 125, 211
Markov model, 93
Medical applications, 9
Medical sensor, 30, 210, 214
Memory, 94, 96, 122, 124, 143, 204, 234
Messaging technology, 51
Microprocessors, 134
Microservice, 59
MLQD, 225, 226, 229
Mobility, 10, 57, 155, 183, 193
Motion sensor, 31, 175
MQTT, 47, 51, 202, 206

N

Name Resolution, 57
NBIoT, 21
Network layer, 3, 5, 6, 23, 46, 48, 128, 158, 211
Network security - 116, 131, 211
Node capture, 115
Noise sense, 61, 62
Non repudiation, 133

O

Oxygen meter, 31

P

PaaS, 36, 141
Perception layer, 5, 6, 23
Position sensor, 26
Processing layer, 6, 200, 203
Proximity sensor, 26
Pulse sensor, 30, 143

Q

Qualifier, 142
QoS, 50, 57, 75, 104, 173, 175, 179, 182, 189, 193
Quick failure recovery, 120

R

Referential IoT platform, 92, 93
Registry, 101, 108, 176
Resilience manager, 58
Respiration rate, 30, 212, 213, 214
RFID, 1, 4, 35, 36, 53, 76, 94
RMI, 47
Routing protocol, 17, 76, 119, 128

S

SA, 99, 109
Scalability, 52, 63, 75, 94, 95, 108, 134, 209, 234
Secure agreement, 40

Selective Forwarding Attack, 129
SensIaaS, 138, 141, 144, 146
Service consumer, 101
Service provider, 47, 101, 133, 176, 178, 179
Short range communication, 26, 32, 76, 94
SigFox, 34
Sink, 4, 139, 198
Skin conductivity, 30
Smart city, water, health, living, 11, 12, 210
Smart management, 168
Smart transportation, home, 10, 161
Sniffing, 66, 125, 211
SOA - 46, 47, 59, 75, 100, 101
SOAP, 47
Software compatibility, 94
SOSA, 101, 102
Sound sensor, 26
SpO2, 30, 142, 146
SQME, 105
SSN, 100, 101, 105, 106, 108, 110
Static node, 39
Stress sensor, 30
Stretch attack, 110, 128, 129
Symptom collection, 146, 147
Synchronises, 204, 217
System privacy, 38

T

Temperature sensor, 26, 31, 146
Traffic surveillance, 10
Trust and governance, 132

U

URN, 4

V

Vampire attack, 127, 128
Vectors, 122, 132
Virtualisation, 36, 58, 78, 118, 126, 144, 150, 200, 201
Virtualised device manager, 57
VM, 78, 124, 125
VSA, 101
VSDN, 57
Vulnerability, 120, 126, 132, 134

W

WAN, 20, 46
WBAN, 22, 32, 33, 38, 212
Wearable devices, sensors, 9, 16, 28, 30, 39, 40
Web API, 47
Web integration, 122
Web services, 63, 100, 109
WSN, 2, 12, 17, 19, 21

X

XML, 47, 62, 209

Z

ZIgBee - 32, 33, 94, 198, 202, 205